Effective Thermal Insulation - The Operative Factor of a Passive Building Model

Effective Thermal Insulation - The Operative Factor of a Passive Building Model

Editor

Teodor Piatek

scitus
academics

Effective Thermal Insulation - The Operative Factor of a Passive Building Model
Edited by **Teodor Piatek**

ISBN: 978-1-68117-524-9
Library of Congress Control Number: 2016942884

SCITUS Academics LLC,
www.scitusacademics.com
Box No. 4766, 616 Corporate Way,
Suite 2, Valley Cottage,
NY 10989

Printed in United States of America on Acid Free Paper

Preface

Thermal insulation is the reduction of heat transfer between objects in thermal contact or possibly in the range of radiative effects. Thermal insulation can be carried out with specially designed processes or methods and with appropriate object forms and materials. The whole building envelope must be well insulated in low-energy buildings in climates with a heating demand. The building envelope is made up of all elements that separate the interior from the exterior. Its main purpose is usually to offer a cozy indoor climate, regardless of the weather-driven outdoor climate. The heat inside the building wrapping is usually above the outside during cold periods. As a result, heat is lost through the wrapper and the interior of the structure cools down to adjust to the external temperature unless this heat is replaced. The reverse applies in warm periods when the building enters through its envelope with excessive heat. Therefore, despite the climate, it makes sense to limit the heat flow in any building – and this is exactly where thermal protection enters.

This book presents wide-ranging theoretical perspectives and real-world applications in the energy saving concept and a solitary action for this category of subjects. The aim of the book is to illustrate the concept of thermal insulation, such as processes, and the attempt to build an efficient passive construction model.

Table of Contents

Chapter 1

Improvement of Thermal Insulation by Environmental Means

Amjad Almusaed[1] and Asaad Almssad[2]

[1]Archcrea Institute, Aarhus

[2]Umea University, Umea

1. INTRODUCTION

Insulation is a vital part of all contemporary buildings; it performs many functions, all of which influence the cost of the building and its operating cost. This component is essential to be positioned not only in the floors, walls, and ceilings of the buildings, but also using of other key technique to improve the insulating process (John F. Malloy 1969). There are many other ways, although one of the most vital way is using of vegetate buildings concept.

Building and garden usually do not arise together and seldom at the same time. Mostly a building is build first and the garden made around it, but if there was a garden first, little of it remains undisturbed by the time the building is build. The layout of the building requires a planner, that of the garden also. Often the representatives of these two entwined disciplines do not meet, or come together to late, when they can but tolerate each other. It would be better if they met to discuss and decide every detail before the first sod was broken. Best of all, the planning of building in its garden should be a mutual undertaking. (Jan Birksted 1999). No Architectural concept is complete without natural areas. Exclusive of soil such growth media to grow plants or vegetations, without water to encourage them, and without the wildlife attracted by the sustenance thus offered, an architectural element has not the fully rounded totality of a factual architecture. The most important class of environment means in reading of this chapter is green areas inside and outside the architectural elements, which requires be to implicit more in terms of ecology as an interface between us and the natural world. Therefore, a green building comes into sight such global human requirement. Today the requirement is to oriented building components towards natural resources to be included in building concept. The green areas is the most significant environmental means, where the green covering concept can be changed to the concept of biophilia.

Biophilic habitat combines the interests of sustainability, environmental consciousness, green areas of the large nature, and organic approaches to evolve design solutions from these requirements and from the characteristics of the site, its neighborhood context, and the local microclimate. The concept of biophilic architecture is a part of a new concept in architecture, that labor rigorous with human health, ecology and sustainability principles, such a integrate part of architectural configuration, which must be in optimal proportion with other buildings area. At what time an architectural element is viewed as an ecosystem, it is obvious that biophilic architecture can play a vital role in creating a healthy indoor environment.(David Pearson, 2004). The biophilic architectural concept deals with the interaction and interrelations of communities of human and plants with under architectural spaces upon local microclimate. A green areas concept can improve the building functions by increasing the efficiency of energy resource, and reducing the building impacts on human health and the environment during the building's lifecycle, through better sitting, design, construction, operation, maintenance, and removal. (Frej, Anne B. 2005).

Energy is fundamental to all life. Even early man knew his life depended upon energy from food, fire, and from the sun, and he conserved it to the best of his ability. He stored food, built a shelter around his fire, and wrapped himself in skins. The shelter around the fire to contain its heat, and the skins wrapped around his body to retard the flow of heat from his skin to the surrounding air were two types of thermal insulation. Therefore, thermal insulation was one of man's first inventions. This illustrates that the need for energy conservation is as old as man himself (John F. Malloy 1969). The new orientation of actually researches on biophilic habitat aims to move the human actions under an architectural roof towards the green of the large nature; this movement intends to create:

- Natural and physical frameworks become more than friendly.
- The Energy consummate by our buildings is most well organized.
- The human development by effectively managing of natural resources is effective.
- The negative effects of climate change become more reduced.

2 ENERGY CONSUMPTION AND MACRO-ENVIRONMENT METROPOLITAN

2.1 The negative effects of global climate change

Throughout mainly of the geological record, the Earth had been bathed in uniform warmth such was the fixed opinion of geologists. The glacial epoch it seemed to have been a relatively stable condition that lasted millions of years. During the last 2 billion years, the Earth's climate has exchanged between a frigid "Ice House", like today's world, and a sweltering "Hot House", like the world of the dinosaurs. Global climate change is reasoned by the accumulation of greenhouse gases in the lower atmosphere. The global concentration of these gases is increasing, mostly unpaid to human activities, such as the combustion of fossil fuels (which release carbon dioxide) and deforestation (because forests remove carbon from the atmosphere), cities extending and wrong consumption of our natural resources. Extreme weather events such as droughts, floods, cyclones and frosts may affect areas previously unaffected or strike with increased frequency. The sun influences Earth's climate (Amjad Almusaed 2010). What is new is that the changes predictable to occur as quickly that nature will have more than tricky to keep up. When the climate revolves out to be warmer, we have to remain for that some species will get it too hot for us, but could flourish further north (McMichael, A. J., and Haines, A. (1997)).

Human beings are exposed to climate change through changing weather patterns (temperature, precipitation, sea-level rise and more frequent extreme events) and indirectly through changes in water, air and food quality and changes in ecosystems, agriculture, industry and settlements and the economy.

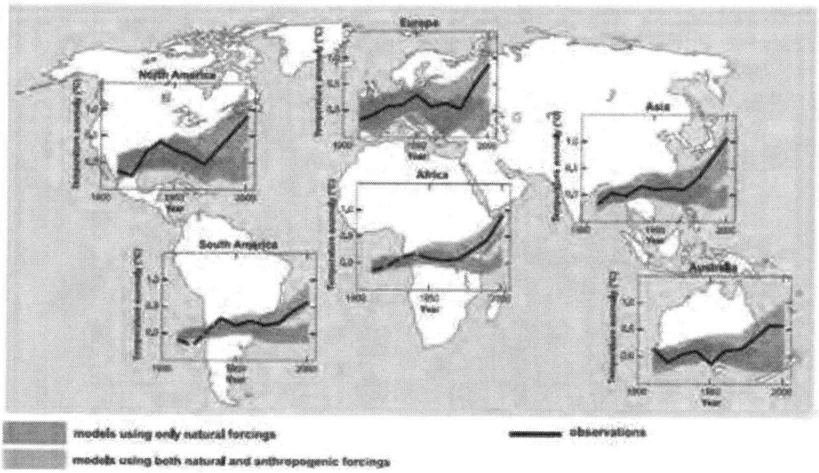

Figure. 1. The global and continental temperature change (Source: IPCC 2007)

2.2 Negative effects of urban head island phenomenon

One of the most important issues facing Biophilic cities of the future is the urban heat island effect, which will be greatly make worsted by rising global warming. The major reason of the urban heat island is change of the land surface by urban progress; waste heat creates by energy usage is a secondary contributor. As inhabitants centers grow they are inclined to adjust a greater and greater area of land and include an equivalent amplify in average temperature. Partially as a result of the urban heat island effect, monthly rainfall is about 28% greater between 30-60 kilometers downwind of cities, compared with upwind. Heat islands can affect communities by increasing summertime peak energy demand, air conditioning costs, air pollution and greenhouse gas emissions, heat-related illness and mortality, and water quality (Amjad Almusaed 2010). They can be developed on urban or rural areas. As it would be predictable, there is a minor fact regarding non-urban heat islands, since they typically do not correspond to a risk for the human being or the environment. In the meantime, urban heat islands have been abundantly addressed throughout decades in urban areas with an extensive variety of climates and landscapes (Amjad Al-musaed 2007).

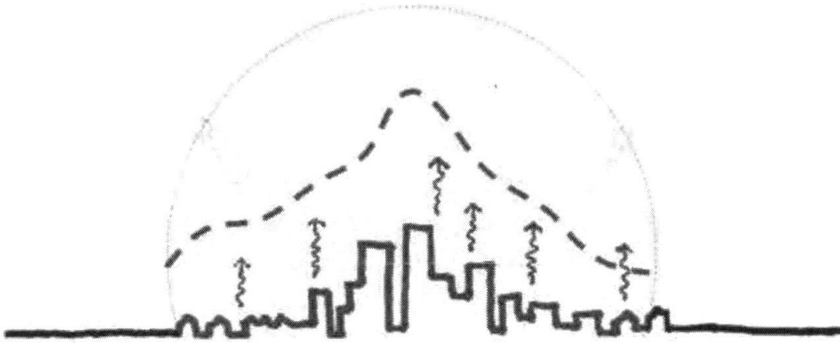

Figure. 2. Urban heat island dealings (sources: Amjad Almusaed 2010)

2.3 Improvement of energetic macro environment

Plants, vegetations upon building surfaces and are a method not only to decrease city temperatures but also to diminish the heating load and energy require of individual buildings. A long-term strategy of planting shade trees and creating of reflective buildings materials for roofs walls, and pavements can mitigate the urban heat island effect and help to diminish associated economic, environmental, and health-related costs. (H.Y. Lee 1993). Green areas supply always the important environmental and human health benefits which cover a large area of advantages and benefits that can be for example in ameliorate the urban island effect in special for hot climates and relieving the damage on the ecology of the city. Principally concerning microclimate, rainwater retention and filtering of airborne pollutant lowering energy expenditures, purifying the air, reducing storm-water runoff, longer durability of the building skin, due to lower surface temperatures and better protection against UV-radiation, creation of recreation areas in parts of the city, aesthetical improvements in denaturalized urban centers and many others. Numerous reimbursements can result from the adoption of green areas over the buildings and using the new concept of biophilic city. Vegetative building exterior skin can also play a vital function in addressing UHI in global cities, as they have been well-documented to decrease building surface temperatures and building heat gain (Liu & Baskaran 2003, Del Barrio 1998) and are rising in status due to their thermal and ecological characteristics.

One introduces additional green areas into the built environment, and the other engages choosing correct building materials that reflect the sun's rays. Both strategies diminish the urban heat island effect - the temperature in center cities is at 2-10 degrees higher than in nearby rural areas. With using of a light-colored building surfaces and materials or reflective coatings lowers surrounding temperatures. These measures may limit the frequency, duration and strength over periods of hot weather. Strategies to reduce overheating, such as the use of cold skin building and clean sidewalks, and planting trees providing shade, have many advantages.

3 ENERGY CONSUMPTION AND THERMAL BUILDINGS MICRO-ENVIRONMENTS

3.1 Reducing of energy consumption upon micro-environment by using green areas

Communities can take a many steps to save energy consumption upon micro-environment. These strategies include: By means of greater, the concept of biophilic urban and architecture, vegetated buildings extern surfaces, by living green walls and planting trees and vegetation employ the evapotranspiration and evaporative-cooling procedures of vegetation on construction surfaces and integrate open green spaces. In addition, trees, shrubs, and other plants help reduce ambient air temperatures during a process known as "evapotranspiration." This happens when water absorbed by vegetation evaporates off of the leaves and surrounding soil to naturally cool the surrounding air. Trees also insert oxygen to the atmosphere, break down a quantity of pollutants and diminish dust (Amjad Al-musaed 2007). It has been predictable that 300 trees can counterbalance the quantity of aerial pollution that a human being generates in a life span. 1 m^2 of green areas can remove up to 2 kg of airborne particulates from the air every year, depending on foliage type.

Reducing the level of heat-absorbing surfaces such as paved, asphalt or concrete surfaces and amplify their permeability, where the certain that the individual built form's configuration (size, clustering and form) does not give confidence heat-island effects (Myer, W. B., 1991).

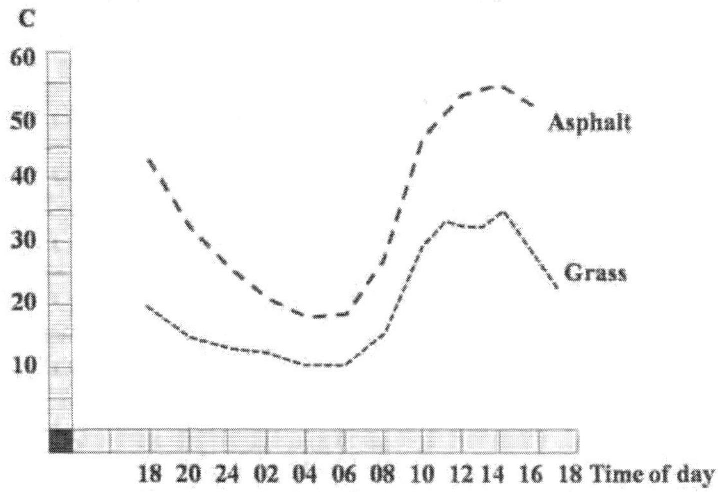

Figure. 3. Earth Surface temperature through 24 hour (Source : Amjad Almusaed 2010)

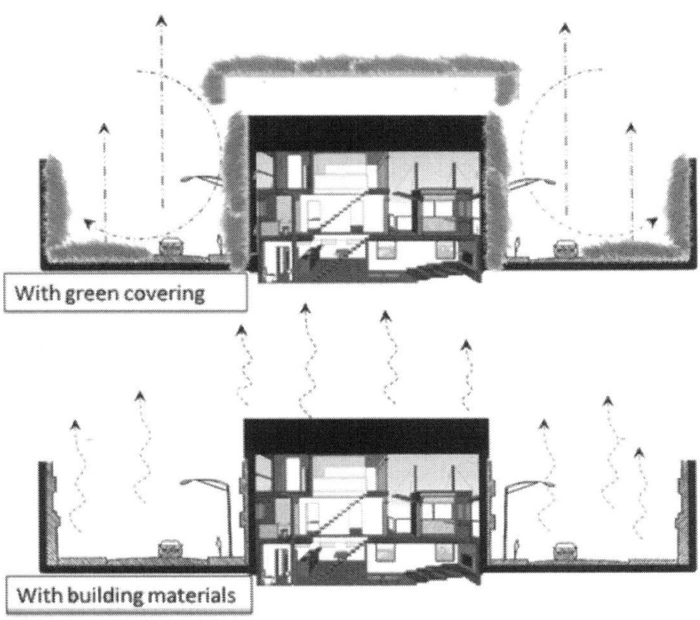

Figure. 4. The comportment of different surfaces (green covering – non green covering)

The replacement of vegetation by streets, buildings and asphalt, frequently guide to a greater absorption of sunlight throughout the day and a slow release of heat throughout the nighttimes. Selection of building material is a key in overturning the heat island effect, for it is the dense dark-colored structures that draw sunlight and keep it for periods. Green walls or roofs, with their landscaping and incorporation of natural materials, are ideal in their resistance to heat absorption. A study by Singapore researchers found that such gardens reduce roof ambient temperature by 4 °C and that heat transfer into the rooms below is lower.

A study in Tokyo shows that if the temperature in Tokyo goes down by 0.8 °C because of rooftop gardens, electric-bill savings equivalent to approximately $ 1.6 million per day could be achieved (Wong Nyuk Hien 2008). The urban heat island mitigation strategies, can support to diminish direct energy utilize in buildings, and if applied on a community-wide basis, can decrease generally ambient air temperature in a specified region (Gallo, K.P.; Tarpley, J.D. 1996).

3.2 Reducing of energy consumption by using of soft cool material buildings

Using soft cool building materials and controlled to cool paving materials. Adjust current and new urban city block layouts and configurations with explain patterns, materials and surfaces that absorb a smaller amount of solar energy.

3.3 Increasing of the shading effects

That can take place by assemblage of physical volumes, or planting trees. Planting shade trees reduces the amount of heat absorbed by buildings by directly shielding them from the sun's rays. A local microclimate can be different from its surroundings by receiving supplementary energy, consequently it is a modest warmer than its surroundings. On the other hand, if it is shaded it could be cooler on average, because it does not acquire the direct heating of the sun. Its humidity may vary; water may have accumulated there production things damper, or there may be a smaller amount water so that it is drier.

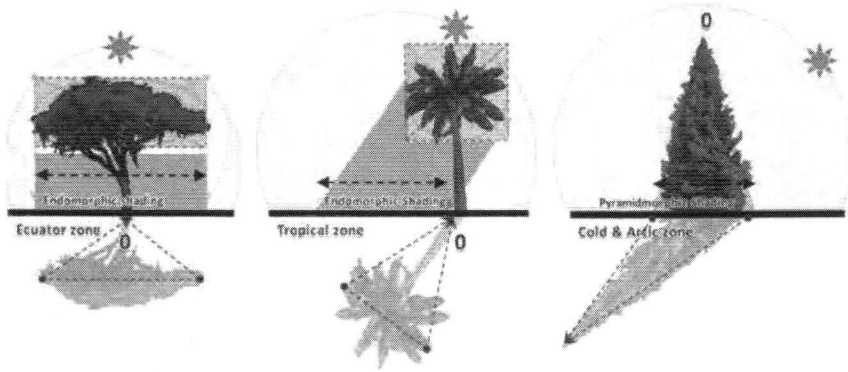

Figure. 5. Trees shade morphologic in correspondence to world climate specific (Archcrea institute)

3.4 Saving energy by using a well reflecting and high building materials

For generate a competent result, of reducing energy consumed by building function, we need to utilize a well reflective and high emissivity building materials for the climatic skin building surfaces or install green areas for the extern roof and facade. Therefore, we require to increasing the reflectivity of buildings surfaces such as rooftops and using frequently of light colors. Creation highly reflective building surfaces will keep buildings cooler and warmer and reduce energy bills. Research conducted in Florida and California indicates that buildings with highly reflective surfaces require up to 40 percent less energy for cooling than buildings covered with darker, less reflective roofs. Opt for roof, surface and building colors so as to decrease effects (evade black or dark colors but utilize white and light colors).

3.5 A well design of circulation arteries

Design the roads and street canyons width, height ratios and their orientations to control the warming up and cooling processes, the thermal and visual comfort conditions, and assist in air-pollution dispersal (Ken Yeang 2006). Design the built form with the topography of the locality, to ensure that the heat-island effect does not affect the climate of the larger region surrounding the designed system and to reduce the wider impacts on people and on the surrounding natural and built environment (Ken Yeang 2006).

3.6 Slow thermal reactions leading to formation of ozone pollution

As a result we require to control the traffic-systems reduction, distraction and rerouting to reduce the production of air pollution, and heat discharges. For parking the optimal solution is in building vehicular parking spaces underground or as covered structured parking. Use an open-grid pavement system (with impervious surfacing such as porous concrete) for the parking-lot areas (Ken Yeang 2006).

3.7 Reducing of the energy consumption

In the past, green areas on the roofs have been used to insulate edifices. The major and vital role of green areas on biophilic architecture is that to conserve, insulate and hold back a change of energy flux, between outside and inside. The green areas amplified the thermal performance of the green covering system and constantly lowered the heat transfer between the construction and its environment all over. Green areas insulate buildings by preventing heat from moving throughout the climatic skin areas.

3.8 Increase of the physical comfort and the quality of the life

The economic price of the success strategies is outweighed not merely by the cooling energy reserves, but in addition by the decrease in greenhouse gas releases, esthetic value of urban forestry, and the increased quality of human health (Hinkel, Kenneth M. (March 2003). These can be defined as win strategies. Mitigation of the urban heat island impact by increasing the employ of surfaces covered in vegetation and building materials with higher than usual reflectivity; in mixture with a strengthening of emissions decreases programs has the potential. Using top roof such climatic skin roof can help our mitigation strategy for reducing of urban heat island effect.

3.9 Reducing of the buildings height

Using of a very high buildings in the centre of the cities increase temperature few degrees. The high buildings surrounded by many urban areas give a multiple surfaces for reflection and absorption of sunlight, increasing the efficiency with which urban areas are heated. This is called the" canyon effect"

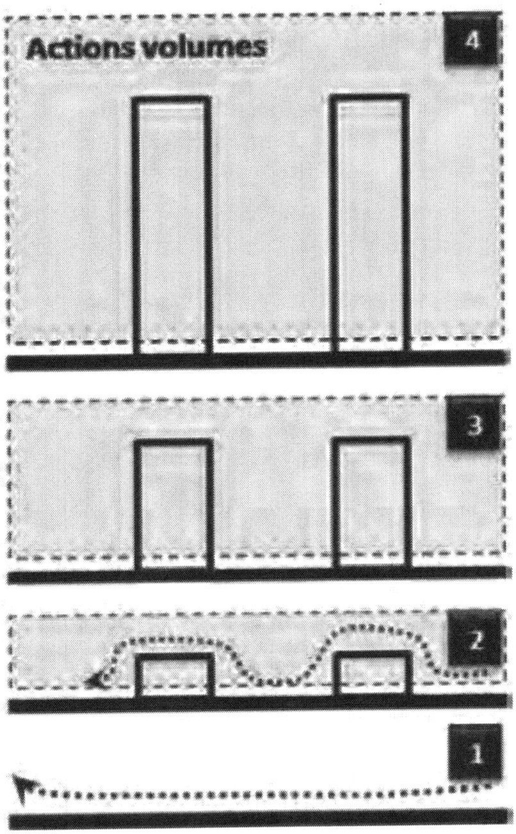

Figure. 6. The effects of high building on city climate

4 THERMAL PERFORMANCE OF GREEN COVERING

Green areas are the most important visual associations between land, buildings and the sky; the most prominent of all plant life, and without their presence, our townscapes would be naked. A sense of continuity is given by old green area and they remain well-known marker when unneeded buildings, hedgerows and path make way for new developments. The green areas amplified the thermal performance of the green covering system and constantly lowered the heat transfer between the construction and its environment all over. It insulates building by preventing heat from moving throughout the climatic skin areas. For cold and temperate climate the energy flux occurs from hot inside spaces to cold outside environment and contrary meant for hot climate. Thermal insulating green area build up with official property values are permitted to be supplementary to the conventional thermal insulation.

Due to this special build-up, the building owner saves approx. 2 litters / m^2 fuel oil per year. The green areas on building surfaces reduced the daily energy demand due to heat flow through the building surfaces by 83-85 % in the spring/summer and 40-44 % in the fall/winter, with an overall annual reduction of 66 %. Green areas insulate buildings by preventing heat from moving through the climatic skin areas. Their insulation properties can be maximized by using an increasing medium with a low soil density and high moisture comfortable and by selecting plants with a high leaf area directory. In the winter, the additional insulation supplied by the growing medium (substrate) helps to diminish the amount of energy necessary to heat the building. The amount of the energy rate savings impact is a function of (Amjad Almusaed 2008):

- The size of the building
- The building location
- The depth of the growing medium
- The type of plants and other variables

Since the 1980s investigate has been conducted on topics such as the insulating effects of greens on façades. Green areas over building surfaces have been shown to significantly reduce building surfaces temperatures and building surfaces heat gains. Karen Liu's field experiments in Ottawa, Canada confirmed that an extensive green area reduced heat gains by 95 percent and reduced heat losses by 26 percent as compared to a standard reference area (Liu & Baskaran 2003). In experiments at Pennsylvania State University (PSU), roof surface temperatures were below ambient air temperatures in greened roof areas at the same times that temperatures on traditional roof surfaces reached 40 degrees Celsius above air temperature. PSU studies also indicated significant (5 – 10 degree C) differences in indoor air temperature in rooms below greened and non-greened roof areas (Gaffin et al 2005).

The vegetative skin building was modeled as three divide layers – the building material layer, the soil surface layer, and the canopy layer. Each is represented by its own energy balance, as seen in the figure below. The associated equations can be linked together by the flux through each connecting boundary. This makes it possible to solve for surface temperature taken at the soil surface, as it is most easily measured in a green roof. Heat flux into the building can also be solved for using these energy balances, as well as water lost through evapotranspiration.

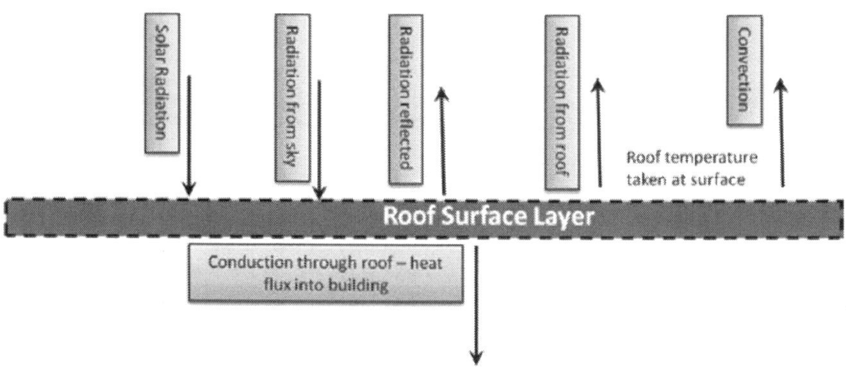

Figure. 7. Energy balance on roof surface layer for typical, reflective, and elastomeric roofing (Source : Caroline H. 2008)

Temperatures of classical building surfaces exceeded ambient temperatures by up to 45 °C and had ranges of skin building temperature also exceeded 45 °C. Vegetative skin buildings maintained temperatures under the ambient during the day in the majority cases, and had average temperature levels that were at or under the ambient environmental temperature (Caroline H. 2008).

The average range in temperature for a green skin building was 10 °C, while the average range in classical skin building was 42 °C. These vast roof temperature ranges can create stress in the structural roofing materials themselves, which is one skin building of the reasons that green skin buildings are able to extend the life of building materials. Vegetative skin building temperatures and fluxes were the lowest category of skinning building condition during the midday hours in the greater part of cases, declining midday temperatures and fluxes would be mainly significant in office buildings that not only have highest solar heating loads at that time, but also highest heat loads from high habitation and equipment operation.

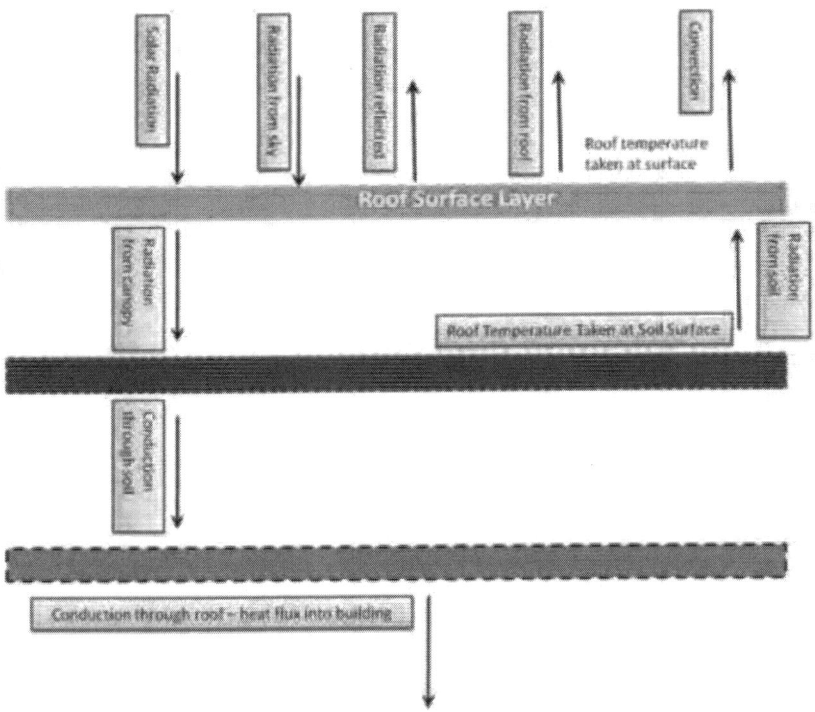

Figure. 8. Energy balance on vegetative skin building layers (Source : Caroline H. 2008)

Additionally, the temperature of green skin building surfaces and the heat flux through the green skin buildings had a lower range of values than any other kind of skin building. While the average values of temperature and flux were often inferior for elastomeric skin buildings, the range of values was much lower for vegetative skin buildings. The elastomeric building kin surfaces much more regularly had negative values of flux (heat loss) mainly in the morning and evening hours. These negative flux values of elastomeric skin buildings were in several cases as large as or larger than the maximum positive flux. The fact that green skin building temperatures and flux values were most stable throughout the day, representing the lowest range from morning to midday, is also considerable for building operators to note. This means that cooling and heating loads will be consistent throughout the day. While any kind of alternative skin building was confirmed to greatly decrease the flux into the building, it should also be an objective to maintain flux and temperature so as to decrease the pressure put on heating and cooling systems to adjust for changing heat fluxes. In particular, the negative heat fluxes that often were demonstrated by elastomeric skin buildings in the model results would indicate that buildings might need heating in the mornings to maintain room temperature due to this heat loss(Caroline H. 2008).

4.1 Greenly areas placement

The green areas can take a differ places in relation to the non- greenly areas where the green area appearance aim to be synchronized by means of other area in concordance with architectural perception upon biophilic habitat. The stabilizer forms resulting from the accumulation of separate elements, which can be characterized by their capability to develop and combine with other forms. For recognize preservative groupings as integrated compositions of shapes as figures in the visual field, the combining elements have to be connected to one another nr a rational method. Good biophilic habitats plan their planting to avoid unfavorable local microclimates avoiding frost pockets for sensitive crops, and allowing for the effect of aspect on temperature or water balance. They can also try to make new microclimates, which will favor the plants they are growing. Shelterbelts of planted trees or bushes create a drag that slows down the drying or cooling winds that blow across architectural volume. The effect of a shelter belt of trees on wind speed can extend across the field as far as 20 to 30 times the height of the plants (Jonathan Adams 2007). By means of the green areas form and position over architectural concept, it can be measured by three criteria:

- Performance.

- Identity.
- Economy of means.

Everyone has a subconscious or usual means to be familiar with the architectural elements that are used every day symbols of comfort, familiar functions and occasionally, visual excitement (Amjad Almusaed 2010).

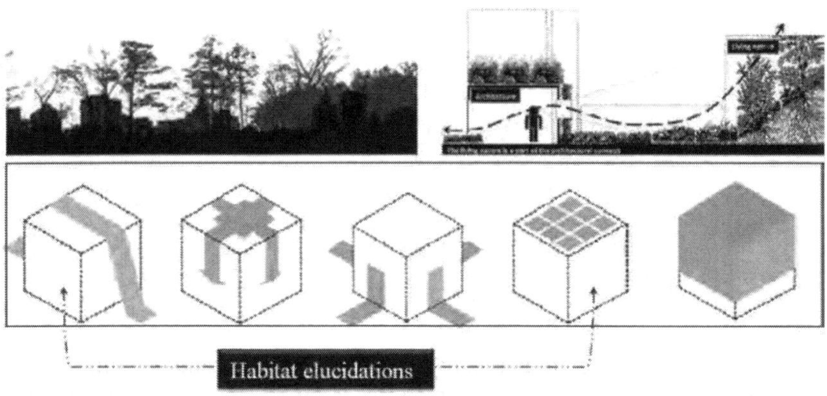

Figure. 9. Architecture and green covering forms and placements upon biophilic habitat (Amjad Almusaed 2010)

4.2 Greenly effects on the environment

The carbon is incorporated into the tree's growth. Because of transpiration and shading, the air surrounding a tree can be around 5 °C cooler than its environment. Tree-shaded neighborhoods can be up to 3.5 °C cooler than those without trees. The competence of plants to produce oxygen varies quite a bit. It is also potential to build an artificial process involving photosynthesis that would successfully do the same thing but it would not be a beautiful to walk through (Ken Yeang 2006). An average of human requires are; 2600 grams of food, 686 grams of oxygen (O2) and 400 grams of water.

4.2.1 Vertical green

A plant leaf produces about 0.005 litter's oxygen per hour. Therefore a mature human need about 50/0.005 = 10000 leaves which would be provided by about 500 small plants for one person.

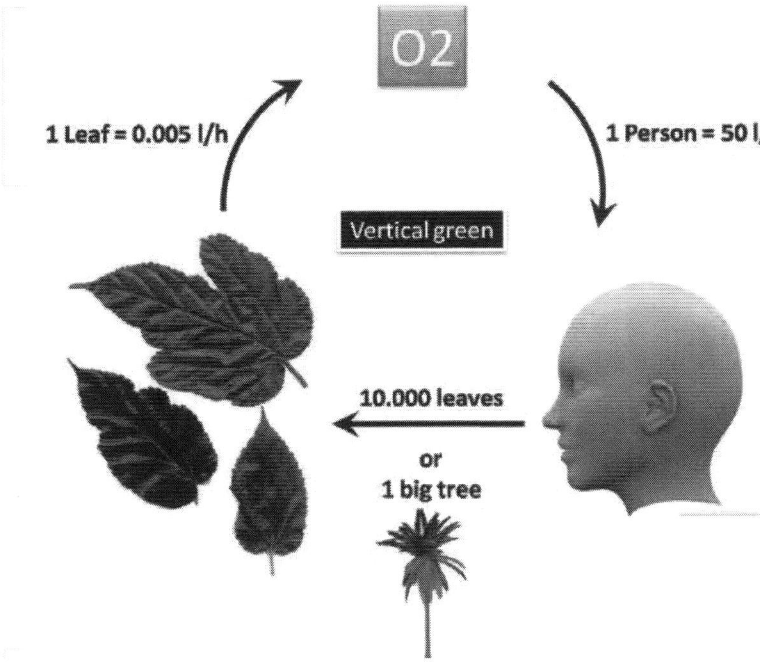

Figure. 10. Human and plant interaction on vertical green

If the average of shrub or other medium size plant has 30 leaves per plant, then that would be 5 ml / leaf x 30 leaves = 150 plants (Wizkid 2008). An average of 18 cm^2, leaf area can release atmosphere of 0.005 litters' oxygen per hour. An average of person who consumes 50 litters oxygen per hour. Consequently, an average of 18 m^2 of vertical green area is sufficient for one person. In addition, an average of 5 m^2 of vertical green areas is satisfactory. There are many assumptions, average leaf, and average plant.

4.2.2 Horizontal green

In a 1.5 m^2 of uncut grass, produces enough oxygen per year to supply one person with their yearly oxygen intake requirement (Brian Burton 2009).

In addition it will necessitate to take into evidence oxygen production reduces as carbon dioxide concentration increases, assuming this hypothetical person is in a limited space with all these plants, the CO2 concentration will increase suitable to the person's expiration. This will slow down the plant's photosynthetic rate (Jonathan Adams 2007).

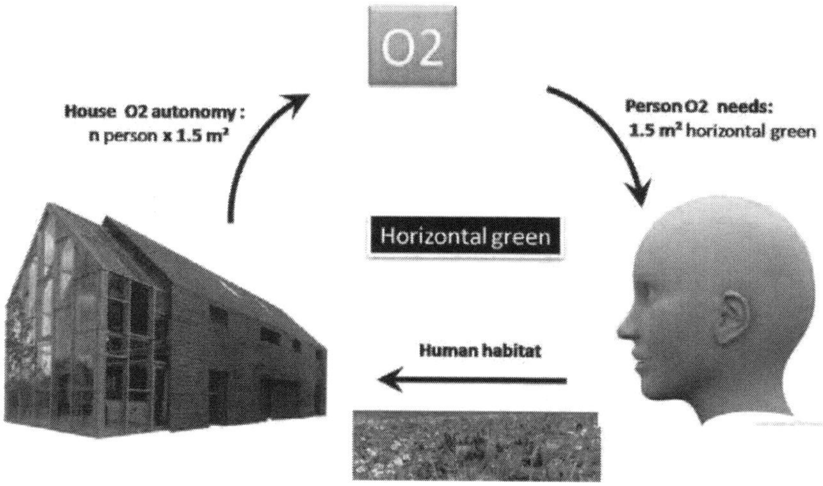

Figure. 11. Human and plant interaction on horizontal green

Hospitals and health facilities utilize the therapeutic benefits of green areas. These facilities sometimes use gardening as a tool to enhance the healing process for patients. In addition, the person can enjoy the comfort, fresh air, and landscape while restoring their health (Ismail Said (Jun 2003)). The query is how we can obtain the oxygen and air quality from the plants. biophilic structure on the earth is a valued and appreciated part of life, where areas and human carrier green is not only an excellent synthesis of both qualitative and quantitative research that documents the bond between people and plants, it is a synthesis of the life's work and thinking of one of the most important figures in people-plant relationships.

Using of a good managed green covering. According to the NASA study, the heat island effect in urban areas can be most effectively reduced with more green space (vegetation offers moisture to cool the air). In adding, light-colored surfaces can reflect sunlight, and should be used on rooftops (J. Hansen, R. et al (2001). Excessive using of solid elements with less thermal properties such as some of building materials in the front of a less using of soft materials with high thermal proprieties such plants amplify the phenomenon radically (Henry J, Glynn, Heinke G 1989).

When green areas are replaced by asphalt and concrete upon roads, buildings, and other structures, it becomes essential to provide accommodation-growing populations. These surfaces absorb - rather than reflect - the sun's heat, causing surface temperatures and overall ambient temperatures to increase see table 1.

The climate performance of the biophilic architecture be able to be considerably affected by green walls, as well the visible changes concerning temperature, the solar gain by direct solar radiation and long-wave heat as well as convection. In addition the, changes in the humidity levels are also supposed. It is essential to image what an exterior living wall will look like during the winter. Green roofs have extended term experimental in value, variety, conception (Velazquez L, Kiers H 2007). All elements can be given differently into all numbers of permutations combinations of solid and almost transparent membranes (Amjad Almusaed 2004).

Green areas are the most important visual associations between land, buildings and the sky, the most prominent of all plant life, and without their presence our townscapes would be naked. Our modern lives seem to be dominated by conflicts of one kind or another, and on the particular subject of trees it is the pressure on land and the rise of consumer power that is placing the professional adviser and his love of green areas in some difficulty. Today modern architecture may fairly be said to have won its first battles all over the world, but in very few of them has it had any assistance from landscape architecture.

One of the best qualities of the modern movement is its increasing awareness of the connection between the space within building and the space around them, and of the interdependence of building and green areas Other environmental mains of thermal insulating

5. OTHER ENVIRONMENTAL MAINS OF THERMAL INSULATING

5.1 Double skin façade (the energetic role of double skin façade)

Walls must give building spaces protection against hot, cold, wind, external noise, and enhance security. A well insulate heavy construction is needed. But also a sustainable external element is necessary. For an efficient bioclimatic building architect can oriented to a curtain wall such as sustainable exterior elements. Curtain wall is synonyms: double skin façade double-leaf façade, double façade, double envelope, wall filter façade, and ventilated façade (Amjad Almusaed 2010).

Table 1. The different categories of green walls and roof (Archcrea institute)

Criterions	Extensive green walls		Intensive green walls		Green roof
	Spots green suspended walls	*Compact green suspended walls*	*Living walls*	*Energetic biophilic walls*	
	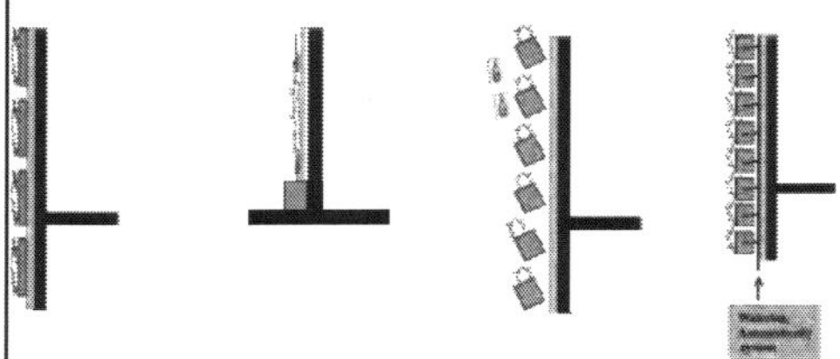				
Thermal insulation	Low, very diminutive	Low, diminutive	Middle, can be efficient	High, more efficient	High, excellent
Environment friendly	Low, is more decorative	Middle, need relationship between building and green areas	Middle, need relationship between building and green areas	High, green is a part of buildings model	High, green is a part of buildings model
Urban heat effects reduction	Low, employment	Middle, employment	Middle, employment	High, effective in combating the phenomenon act	High, effective in combating the phenomenon act

The curtain wall on façade is principally a couple skins separated by an "air corridor". The main layer of skins is usually insulating. The air space between the skins layer is as insulation against temperature extremes, winds, and the sound. If there are two skins of glass, or other thermal opaque materials so for shading interior space that the sun-shading devices are often located between the

The double skin façade consists of two layers of materials, with air space between the two layers preserved, the principal's roles of curtain walls are controlling of solar gain, access to fresh air, embodied energy, esthetics. see figure 12.

Of course, there is a certain level of energy consumption, but this is significantly reduced as internal temperatures, are already lower than outside temperatures. Double –skin facades offer a protected from the exterior environmental conditions; these shading devices are less expensive than system mounted on the exterior.

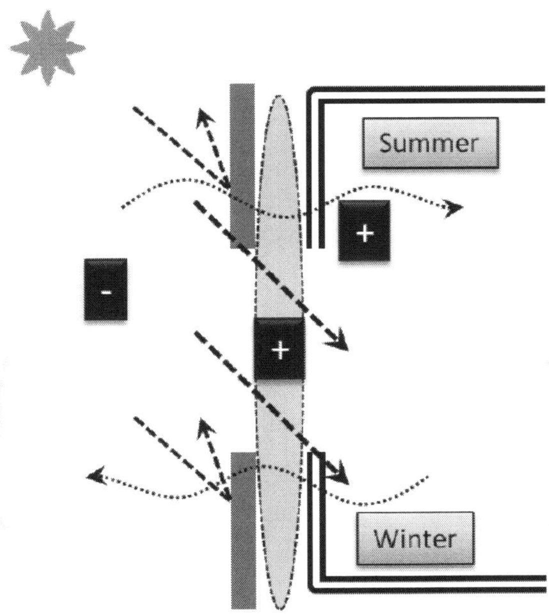

Figure. 12. The thermal role of curtain wall

The principal benefit of double-skin façades over traditional architecture is that they permit the application of blinds even for the buildings with substantial wind. The special materials are mainly used for architectural purposes due to their performance in reducing solar heat.

Energy saving is an essential factor to reduce the emission of carbon dioxide which is a cause of global warming, and ventilation is only unique method to control the indoor air quality and also regarded as an effective method to sweep out the indoor.

In one recently experiment, achieved by department of civil engineering and architecture in University of the Ryukus in Japan, by a group of engineers and architects, about the ventilation of living spaces. That was by using two types of passive cooling (one is a conventional building with cross-ventilation through open windows, and the other is a nuilding which has double skin walls and the gaps of these skins are ventilated .The ventilation airflow comes from the gaps between the double skins and is discharged from an exhaust tower on the roof top. It called ventilation model. The latter is an ordinary hours. The outer walls are made from reinforced concrete and the inner walls are made from heat insulation boards, whose material is foamed polystyrene, and plywood. The difference between two experiment models is only the gap between the outer and the inner skins. The resulted of this experiments shows that the indoor air temperature in the ventilation model is lower than the conventional model by 2 degrees in summer and 1 degree in winter except 5 hours in the morning (Oesterle, Lieb, Lutz, Heusler 1999).

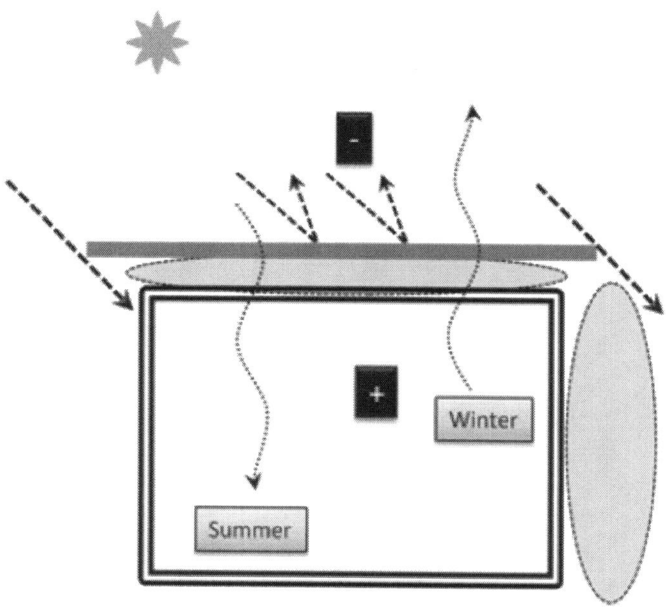

Figure. 13. The double skin roof positive effect

It is clear that the double skin walls of the ventilation model have effects on indoor thermal environment, which means that the double skin walls can keep the indoor space a little cooler all the year round. Several hours in the morning, this relation reverses, because the outdoor air temperature rises quickly and the ventilation model is directly influenced by the outdoor air. It is vital to make clear that the double skin wall can only cool the air inside the structure by several degrees lower than the actual external temperature (Amjad Almusaed 2004). It would be utopian to expect that these systems could provide the same cooling action as air conditioners. On the other hand, integration this construction form into a big intelligent bioclimatic system can give us a better resultant, which we will see that in other parts of research.

The roof on the hot climate houses receives the highest proportion of solar radiation and is also the surface barest to the clear cold night sky. To limit the heat gain, the most effective method is to shade or construct a second roof over the first see figure.

The outer roof will reach a high temperature and it is therefore imperative to separate it from the inner roof, to provide for the dispersion of heat from the space between the tow and to use a reflective surface on them both. The surface of the lower roof should reflect the low temperature heat and for the outer roof a white surface is best.

5.2 Heat break transfer concept

We all depend on energy to get better our lives. But using energy means nothing on its own; it just a way to achieve something else. And we are becoming more aware of some of the problems that come from wasting energy. The significant way of a wasting energy is energy losses by exchange of energy through external elements. Energy losses in a building mainly occur by conduction through external surfaces radiation, and convection. Conduction takes place when a temperature gradient exists in a solid medium, such external wall, windows, roofs, floors. Energy is transferred from the more energetic to the less energetic molecules when neighboring molecules in collide. Conductive heat flow must occur in the direction of decreasing temperature because higher temperatures are associated with higher molecular energies. Heat transfer through radiation takes place in the form of electromagnetic waves, mainly in the infrared region. The radiation is emitted by some body as a consequence of the thermal agitation of its composing molecules. In a first approach the radiation is described for the case that emitting body is a so called black body. Heat energy transferred between a surface and a moving fluid at different temperatures is known as convection. Condensation on the windows may be a sign of heat loss.

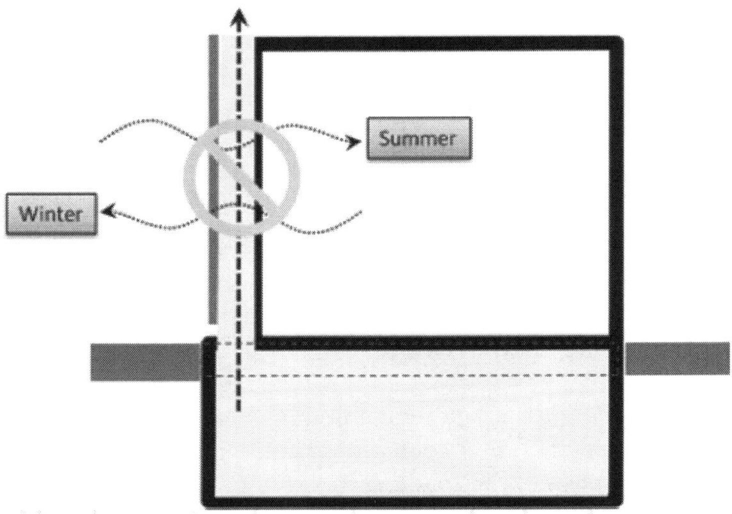

Figure. 14. The constructive effect of the optimistic underground temperature

A damp area around the window from the exterior is another sign of heat loss. During the winter, a typical window loses up to 10 times more heat than an equivalent area of an outside wall or roof. Windows can account for up to 30 % of the heat loss from a conventional house, adding significantly to heat cost. Drafts, window condensation and mould can also affect our comfort and indoor air quality. Sustainability is a wise approach to the way we live. And using energy in a more sustainable way is a part of this approach (Amjad Almusaed 2004, p187). We can save money, reduce imports, protect the environment, and move society forward in an intelligent manner. If we start doing this now we win as individuals and we will win as a society. The concept of intelligent energy losses break consists of a using of some thermal effects to wipe out or stopping the immigration of energy between exterior spaces and interior through external element. This concept can be useful using in architecture on the extreme climate regions. By a deep study of specialists about the optimal thermal effect that can help in realization this concept, consequently we must seek for a suitable source of energy, which must be permanent and easy to get. Creation of this system subsequent to the passive and zero energy concept need a well integrates of the energy in the house's components (Amjad Almusaed 2004).

6. CONCLUSION

A green building is a confusing expression of biophilic architecture. Green building is a construction, which can be shaped by mains of renovation process. While, a biophilic architecture strugglers the negative effects of urban heat island in local microclimate scale, and improves the human physical comfort to create a healthy human life. Therefore, one of the major problems facing us is how to establish and maintain environments that support human health and at the same time are ecologically sustainable. Green areas seems too important to people. Most people today believe that the green world is beautiful.

In fact, green areas by now contribute, some extent, to a better microclimate through evaporation, filtering of dust from the air and reduce in temperatures at the buildings surface. Besides improving the microclimate and the indoor climate, the retention of rainwater is another important advantage. Aesthetic form require, escalating the value of the possessions and the marketability of the building as a complete, mainly for accessible green areas.

On arid climates when the sun rises up, buildings roofs and asphalt road surface temperatures can rise up to 30–45 °C hotter than the air, while shaded or moist surfaces frequently in more rural environs remain close to air temperatures. City surfaces with plants offer high moisture levels that cool the air when the moisture evaporates from soil and plants (Parker, David E. 2004). The influence of plants must employment eventually to keep up with the increased require in energy. Improving energy efficiency can decrease the global warming effects of carbon in the atmosphere, improving air and water quality, and encouraging sustainable development in the cities. The physical frameworks of the city extends unprompted; consequently, it turns out to be a major area of the city centre. The green areas diminutive and take a negligible part of the city typically marginal. Many fixed edifices (civil and industrial buildings) and mobile elements such as cars, public transport and another feature contributing to the warm cities that will increase the phenomenon dramatically.

One of the most significant subjects for our study is to show how we can discover the best possible manner to realize our earth greener, sustainable, and our buildings agreeable and saves more energy, to help the human to live in healthy and economically framework.

REFERENCES

1. Amjad Almusaed 2010, Biophilic and bioclimatic Architecture, Analytical therapy for the next generation of passive sustainable architecture, Springer Verlag London, UK
2. Amjad Almusaed 2008, Towards a zero energy house strategy fitting for south Iraq climate PLEA 2008 – 25th Conference on Passive and Low Energy Architecture, Dublin, 22nd to 24th October 2008
3. Amjad Almusaed 2007. Heat Island Effects upon the Human Life on the City of Basrah, Building low energy cooling and advanced ventilation technologies the 21st century, PALENC 2007, The 28th AIVC Conference, Crete island , Greece.
4. Amjad Almusaed 2004. Intelligent sustainable strategies upon passive bioclimatic houses, The architect school of architecture in Aarhus, Denmark. Pp203-230.
5. Berdahl P. and S. Bretz. 1997. Preliminary survey of the solar reflectance of cool roofing materials. Energy and Buildings 25:149-158
6. Brian Burton 2009, Green Roofs and Brighter Futures, http:// www. newcolonist.com/greenroofs.htmlBowler P.J. 2003. Evolution: the history of an idea. California. p10
7. Caroline H. 2008, Modeling thermal performance of green roofs, Ecocity World Summit 2008 Proceedings, Yale College
8. David Pearson, 2004. The Gaia natural house book, creating a healthy and ecologically sound home Gaia books limited, UK.

9. Del Barrio, EP, 1998, Analysis of the green roofs cooling potential in buildings, Energy and Buildings 27(2), pp. 179-193.

10. Frej, Anne B, 2005. Green Office Buildings: A Practical Guide to Development. Washington, D.C.: ULI--The Urban Land Institute.

11. Gaffin, S, Et al 2006, 'Quantifying evaporative cooling from green roofs and comparison to other land surfaces.' Proceedings of the 4th annual Greening Rooftops for Sustainable Communities Conference. 11-12 May 2006, Boston

12. Gallo, K.P.; Tarpley, J.D. 1996.The comparison of vegetation index and surface temperature composites of urban heat-island analysis. Int. J. Remote Sens.17, 3071-3076.

13. H.-Y. Lee 1993. "An application of NOAA AVHRR thermal data to the study or urban heat islands". Atmospheric Environment 27B

14. Henry J, Glynn, Heinke G 1989. Environmental Science and Engineering. Prentice Hall, Eaglewood Cliffs, N. J. 07632.

15. Hinkel, Kenneth M. (March 2003), "Barrow Urban Heat Island Study". Department of Geography, University of Cincinnati.

16. IPCC, 2007: Climate Change 2007: The Physical Science Basis. Contribution of Working Group I to the Fourth Assessment Report of the Intergovernmental Panel on Climate Change [Solomon, S., D. Qin, M. Manning (eds.)].

17. Ismail Said (Jun 2003), Therapeutic effects of garden: preference of ill children towards garden over ward in Malaysian hospital environment, Universiti Teknologi Malaysia, Jurnal Teknologi, 38(B) Jun. 2003: 55–68

18. J. Hansen, R. et al (2001), A closer look at United States and global surface temperature change. J. Geophys. Res., 106, 23947-23963

19. Jan Birksted, relation architecture to landscape, E& FN SPON, 1999, London England. Pg 179

20. John F. Malloy 1969, Thermal insulation, Van Nostrand-Reinhold, the University of Michigan, USA

21. Jonathan Adams 2007, Vegetation-Climate Interaction, Springer in association with Praxis Publication, 2007, New Gersy, USA

22. Ken Yeang, 2006 A manual for ecological design, Wiley – Academy, UK

23. Liu, K and B Baskaran2003,.Thermal performance of green roofs through field evaluation. Proceedings for the First

24. McMichael, A. J., and Haines, A. (1997). "Global Climate Change: The Potential Effects onHealth." British Medical Journal 315.

25. Myer, W. B., 1991, Urban heat island and urban health: Early American perspective, Professional Geographer, 43 No. 1, 38-48. North American Green Roof Infrastructure Conference, pp 1-10

26. Oesterle, Lieb, Lutz, Heusler 1999, Double–skin façade, Integrated planning, Prestel, 87- 91Olivia Nugent (April 2004), Primer on Climate Change and Human Health,edited by Randee Holmes

27. Velazquez L, Kiers H, 2007. Hot Trends in Design: Chic Sustainability, unique driving factors & boutique Green roofs. Proc. 5th Annual Greening rooftops for Sustainable Communities Conference, Minneapolis

28. Weng, Q.; Yang, S. 2004. Managing the adverse thermal effects of urban development in a densely populated Chinese city. J. Environ. Manage.70

29. Wizkid 2008, Plants making oxygen, USA state, energy department, Biology Archive http://www.newton.dep.anl.gov/newton/askasci/1993/biology/bio027.htm

30. Wong Nyuk Hien 2008, Urban Heat Island Effect: Sinking the Heat, innovation the magazine of research and technology, vol. 9, Nr.1

Chapter 2

3D Visualization of thermal resistance and condensation problems using infrared thermography for building energy diagnostics

Youngjib Ham and Mani Golparvar-Fard

Department of Civil and Environmental Engineering, University of Illinois at Urbana-Champaign

Department of Civil and Environmental Engineering, Department of Computer Science, University of Illinois at Urbana-Champaign

ABSTRACT

Background

Building deteriorations instigated by material degradations or moisture intrusions are the primary causes for energy inefficiency in many existing buildings. For choosing appropriate retrofits, it is important to carefully diagnose and analyze building areas in need of improvements. In addition to reliable sensing and analysis of as-is energy performance, an intuitive recording and visualization of energy diagnostic outcomes are also critical to effectively illustrate the as-is building conditions to homeowners during retrofit decision-making processes.

Method

Toward this goal, this paper presents a thermography-based method to visualize the actual thermal resistance and condensation problems in 3D while taking static occlusions into account. First, several overlapping digital and thermal images are collected from the building areas under inspection. Using a computer vision method– consisting of image-based 3D point cloud and mesh modeling algorithmsactual 3D spatio-thermal actual 3D spatio-thermal models are generated where surface temperature can be queried at the level of 3D points. Based on the resulting 3D spatio-thermal models and by measuring the reflected and dew point temperatures, the actual R-values of building assemblies are calculated, and the condensation issues are analyzed. Taking static occlusions into account, (1) the distribution of the actual thermal resistance over each building assembly, (2) the detected building areas with condensation problems, and (3) the corresponding geometrical and thermal characteristics are jointly visualized within a 3D environment.

Results

To validate the method and investigate the perceived benefits, four experiments have been conducted in existing buildings. Surveys are also conducted by professional energy auditors. The proposed method provides 3D visual representation of the actual thermal resistance distributions and building areas associated with condensation issues at the level of 3D points across geometrical forms while taking static occlusions into account.

Conclusions

The experimental results and the feedback received from the professionals show the promise of the proposed method in facilitating systematic post-examination of building deteriorations and support retrofit decision-makings. Ultimately, converting surface temperature data obtained from an IR camera into 3D visualization of energy performance metrics and possible condensation problems enables practitioners to better understand the as-is building conditions.

KEYWORDS

Image-based 3D Reconstruction, Building condition assessment, Building condensation problem, Thermal resistance, Building energy diagnostics

1. INTRODUCTION

The emerging energy crisis of the building sector and the legislative measures on improving energy efficiency are steering the construction industry towards adopting new design concepts and construction methods that can decrease the overall energy loads. For example, in order to lower the energy cost required for space conditioning, building practitioners are embracing the concept of net-zero passive houses at the design phase. The use of building materials with higher thermal resistance is recommended for better insulation at the construction phase.

The problem of energy inefficiency is however not limited to the design and construction of new buildings. According to a recent report by the U.S. Department of Energy (U.S. DOE [2010]), around 35% of input energy in existing buildings in the U.S. is still being wasted. A primary source of these inefficiencies is thermal performance problems in the building envelope, which can cause an excess in energy consumption for space heating and cooling purposes. Roth et al. ([2005]) reported that such faulty behaviors can account for up to 10% of the total energy consumption in existing buildings during their operational phase. Despite their significance, identifying and assessing building performance problems and improving the energy efficiency under such conditions is not trivial.

Today, building practitioners widely use Infrared (IR) thermography to identify sources of energy loss in building envelopes. The IR thermography –which detects and measures heat variations from surfaces in a non-destructive manner– is particularly known for its usefulness for identifying thermal defects. Nonetheless, current thermographic inspection processes have several inefficiencies. For example, thermal images captured using consumer-level IR thermal cameras typically have low spatial resolutions (160 × 120 or 320 × 240 pixel) and small fields-of-view compared to their digital counterparts (>1 megapixel). As a result, sensing thermal performance data for buildings, even a single room, requires large numbers of thermal images (e.g., ~300 to 400 thermal images for around $4m \times$

5*m* room based on experiments). Considering significantly large collections of 2D thermal images that need to be manually analyzed for the purpose of assessing thermal conditions, their direct application for practical building diagnostics can be time-consuming and labor-intensive. In order to improve the process of sensing thermal performance using raw 2D thermal images, several studies (Borrmann et al. [2013]; [Ham and Golparvar-Fard 2013]; Lagüela et al. [2013]; Wang et al. [2013]) have recently focused on developing methods for 3D thermal modeling of building environments by leveraging digital photogrammetry or laser scanning, and promising results have been reported.

These methods for 3D thermal modeling provide a detailed representation of the actual temperature deviations on building surfaces, which is very effective for energy diagnostics. Nevertheless, the problem of characterizing and visualizing performance metrics and thermal problems from these models for post-inspection is still unexplored. The analysis of abnormal thermal regions – either from 2D thermal images or 3D thermal models – is still primarily based on the interpretation of the measured surface temperatures such as hot or cold spots. Yet, the surface temperatures alone do not directly represent the energy performance problems. Due to a lack of a benchmark, the interpretation of energy performance problems is still primarily based on the auditors' experiences and knowledge, which can cause the outcomes to be subjective or inconsistent. Considering a limited number of well-trained professionals, it is also likely that auditors' knowledge and experience would adversely impact the quality of the inspection.

Another important issue – which is often not considered– is that the deterioration rate of building assemblies typically vary even over small surface areas. For instance, when it comes to building envelopes, various regions across the surface may suffer from different levels of degradation. As such, their conditions would change differently over time. Therefore, a method for building diagnostics should assess and visualize the as-is building conditions at point-level across geometrical forms of the building elements so that defects can be precisely detected and localized.

The presence of static occlusions in existing buildings has also raised concerns about the applicability of thermography-based methods for building diagnostics. Ideally, for accurate measurement of surface thermal performance, any objects around the building interior surfaces need to be removed prior to thermographic inspections. However, in practice, for most in-use buildings, it is not trivial to take away all objects that block the line of sight due to the immobility of objects or space limitations. Theoretically, there is no way to measure the thermal performance for the building areas that are occluded by objects, except for encouraging practitioners to remove as many objects as possible before thermographic inspection. Thus, this issue remains as an open challenge.

To address these challenges, this paper presents a new interactive method for 3D visualization of the actual R-value of building assemblies and condensation problems. The proposed method builds upon the authors' prior computer vision research on 3D spatio-thermal modeling using large numbers of unordered thermal imagery collected by a consumer-level IR camera [(Ham and Golparvar-Fard 2013)]. Beyond synthesizing raw 2D surface temperature data, the method extracts and visualizes building energy performance indicators from the collected thermal imagery. It facilitates the process of identifying potential performance problems by allowing auditors to spend less time on analyzing large amounts of thermal images, and instead focus on investigating the causes of the problems and analyzing various retrofit alternatives.

The primary technical contributions of this paper are two-fold: (1) a method for 3D visualization of energy performance metrics and moisture issues at the level of 3D points across geometrical forms of building elements; and (2) a method for minimizing the adverse impact of static occlusions on thermographic inspection in real-world building environments. In the following sections, previous works on measuring R-values of building materials using thermal imagery and analyzing condensation problems are reviewed, and their challenges in applying to building condition assessment are discussed. Next, the research objective, underlying environmental assumptions, and the proposed visualization method are presented in detail. Finally, experimental results on four building assemblies in two case studies and the responses from domain expert evaluations are discussed.

2. BACKGROUND

2.1 Thermal resistance (R-value) of building assemblies

Analyzing heat transfer conditions through building envelopes is now a required step in rating the energy performance levels of existing buildings (Fokaides and Kalogirou [2011]). This is because during the operational phase, the thermal resistance of building materials gradually decreases. Thus, the actual R-value of the building assemblies with thermal defects is typically lower than the notional value declared by their manufactures. Declining R-values of the building materials results in unnecessary heat transfer through building façades. This in turn increases the operational frequency of heating and cooling systems for space conditioning purposes. Hence, beyond using building materials with high R-value during the design and construction, it is important to explore how the as-is thermal resistance of building assemblies lowers over time.

Recently, several research groups have focused on the problem of non-destructive measurement of the actual heat transfer conditions of building assemblies using thermal images. Madding's work (Madding [2008]) is one of the earliest studies that propose a method for analyzing thermal images to estimate the actual heat transfer conditions for building envelopes. Based on the environmental assumption of a steady-state heat transfer condition of building environments, this work measured the R-value of drywall assemblies using the indoor surface temperature data obtained from thermal images. The measurements were experimentally conducted in controlled lab environments, and then they were compared with the theoretical expected properties. Likewise, based on the similar environmental assumptions, methods proposed in (Albatici and Tonelli [2010]; Dall'O' et al. [2013]; Fokaides and Kalogirou [2011]) measured the overall heat transfer coefficient of the building assemblies. For validation, several experiments were conducted for a few standard wall assemblies. In these studies, the difference between the measured values using thermal images and the notional values was reported to be in the range of ~10% (Madding [2008]), 10-20% (Fokaides and Kalogirou [2011]), and around 15% for cavity walls (Dall'O' et al. [2013]).

For quantifying the as-is thermal resistance of building assemblies, the current non-destructive method regulated by the international standard is to position heat flux meters (HFMs) –a transducer that produce an electrical signal proportional to the heat rate– on the representative building areas and directly measure the heat flow (International Organization for Standardization (ISO) [2007]). However, considering possible variations of the thermal conditions even over small surface areas, it is difficult to ensure that the thermal resistance measured by using a few HFMs will actually represent the entirety of an existing building envelope (Albatici and Tonelli [2010]; Dall'O' et al. [2013]). Interestingly, studies such as (Albatici and Tonelli [2010]) and (Dall'O' et al. [2013]) report 40-80% and 40-175% error between the measurements of the overall heat transfer coefficient using HFMs and their theoretical expected values. Although these observations are primarily based on limited experiments, when one considers the significance of these errors, the applicability of thermal imagery for quantifying the actual heat transfer condition of building assemblies becomes more attractive.

Despite such promising benefits of using thermography, the direct application of 2D thermal imagery as used in prior works is still challenging. In prior studies, a single or at most a few temperature data were extracted from the designated areas captured in 2D thermal imagery, and then a single measurement of the actual heat transfer condition was performed for each building assembly. Hence, these methods assumed that the obtained thermal resistance is representative of all spots on the inspected building assemblies. However, because of different deterioration rates, the actual heat transfer conditions may vary over small scales across geometrical forms. Thus, single measurements will not represent the dynamic variations in the actual R-values that are typically caused by non-uniform deteriorations in the building materials. More importantly, interactive 3D visualization of these dynamic variations in the context of the building under inspection will provide an opportunity for practitioners to better understand the as-is heat transfer conditions.

2.2 Condensation problems in building environments

In building environments, water vapor condensation happens wherever exposed building surfaces are at a lower temperature than the dew point. Typically, it is not easy to detect small amounts of the condensed water on building surfaces with the naked eye. Failure in detecting the onset of such condensation issues causes the condensed moisture to accumulate over time. Then, molds and dust mites may build up, which ultimately has potential for causing allergic reactions among the building occupants. In addition, the resulting wet surface may cause flaking paints and peeling wallpapers. This is when the building occupants are more likely to detect the problems with their naked eyes. In building environments, windowsills or thermal bridges around corners of the walls typically have low temperatures. Thus, energy auditors need to proactively analyze these areas whether the surface temperature has reached to the dew point.

Several studies have focused on analyzing the condensation phenomenon on building façades and the risk associated with the occurrence. Bellia and Minichiello analyzed the thermal and moisture performance of building assemblies, and focused on surface condensation issues of building façades and interstitial condensations within multi-layer walls (Bellia and Minichiello [2003]). Aelenei and Henriques investigated the external environmental conditions that cause condensation problems in building environments (Aelenei and Henriques [2008]). They concluded that the thermal convection effect and the amount of moisture in the air are the most influential factors in forming condensation problems on building façades. These prior works primarily examined the sources that cause condensation and identified ways that can reduce the risk associated with these problems. Nonetheless, devising energy auditing methods that can easily detect condensation problems in building environments is still an open research area. Today, by comparing the surface and dew point temperature, thermal cameras enable 2D detection of those building areas that suffer from condensation issues (Hoff [2011]). These detections are typically done on a single image displayed on the screen of the thermal camera. Because these images are not geo-tagged, it is often not easy to figure out at a later stage which building areas are associated with the detected condensation problems. Moreover, these 2D assessments do not provide any information about the surface area of the detected condensation problems. Hence, it is difficult at a later stage to quantitatively estimate the significance of their impact.

2.3 Overview of the proposed method

This paper presents and validates a 3D visualization method for quantitative assessment of the as-is building conditions. The method leverages a paired collection of digital and thermal images taken using a single IR thermal camera from the building environments under inspection, and accounts for static occlusions in the building environments. The hypothesis is that the quality of the current building condition assessment procedures can be improved with intuitive 3D visualizations that enable practitioners to (1) explore the distribution of the actual heat transfer conditions for building assemblies; and (2) detect condensation problems in 3D. The overall workflow is illustrated in Figure 1.

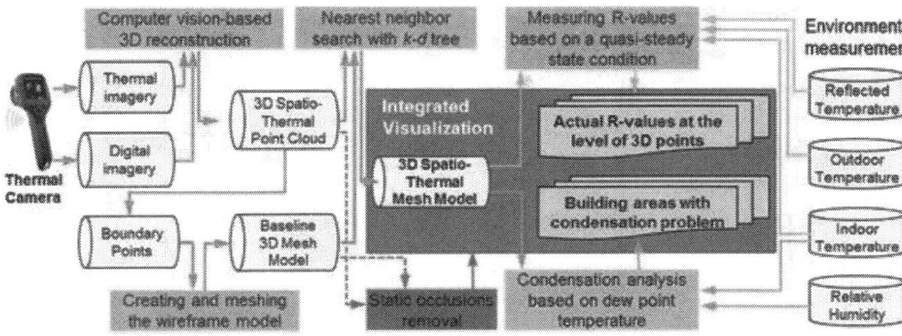

Figure 1. An overview of the data and process in the proposed 3D visualization method for building condition assessment using infrared thermography.

By using digital and thermal images, the proposed method first generates the 3D spatio-thermal mesh models which contain the surface temperature readings at the level of 3D vertices. Then, with environmental measurements, actual thermal resistances are calculated, and building surface areas suffering from condensation problems are detected at the level of 3D vertices in the mesh. Finally, taking possible static occlusions into account, the distribution of the actual R-value, the detected building areas with condensation problems, and the corresponding geometrical and thermal characteristics are interactively visualized within a common 3D environment. In the following sections, the underlying environmental assumptions, experimental setups, and each step of the proposed method are presented in detail.

2.4 Environmental assumptions and experimental setups

The proposed method is based on the following environmental assumptions: (1) the indoor building environment is assumed to have a quasi-steady-state condition of heat transfer during the thermographic inspection process. This is consistent with all recent works on non-destructive measurement of thermal properties (Albatici and Tonelli [2010]; Dall'O' et al. [2013]; Fokaides and Kalogirou [2011]; Ham and Golparvar-Fard [2014]; Madding [2008]). For achieving a quasi-steady-state condition of heat transfer, the building envelopes were not exposed to direct solar radiations and wind loadings as the inspections were conducted before sunrise on a non-windy and cloudy day. This minimizes the influence of the convective heat loss and the surface temperature rise on the exterior of building envelopes. As such, the possible increase in building surface temperatures caused by the release of the absorbed solar energy throughout a day was minimized; 2) it was further assumed that the heat transfer between the indoor surfaces and the thermal camera lens is due to thermal convection and radiation. This assumption is consistent with prior works (Fokaides and Kalogirou [2011]; Ham and Golparvar-Fard [2014]; Madding [2008]). In addition, the building interior spaces were heated for a few hours prior to thermographic inspections, which allowed to clearly capture the as-is heat transfer conditions through building façades during the winter season.

For testing the proposed 3D visualization and conducting professional surveys, several experiments were conducted on four wall assemblies and windows adjacent to the exterior in an existing residential building (Case #1) and an instructional facility (Case #2). For 3D spatio-thermal modeling, digital and thermal images were captured by using a FLIR E60 thermal camera that has a built-in digital camera. Each pair of digital and thermal images were captured simultaneously. The technical specification of the thermal camera used in experiments is presented in Table 1. When temperature deviations used for calculating thermal resistances (Measuring actual R-values at the level of 3D points) are too small, accurate measurement using inexpensive consumer-level IR cameras would be challenging. To alleviate this, similar to (Madding [2008]), the impact of the systematic error on the measurement accuracy was minimized by using a single device for measuring temperature differences.

Table 1. Technical specification of the thermal camera used for data collection

Camera technical items	Values
Built-in digital camera resolution	2048 × 1536 pixels
IR resolution	320 × 240 pixels
Thermal sensitivity	< 0.05°C
NETD	50 mK
Measurement accuracy	2°C or ±2% of reading

3. METHOD

3.1 Reconstructing 3D building thermal performance using 2D thermography

The first step to assess and visualize as-is building conditions at the level of 3D points is to model the actual thermal performance of the building under inspection in 3D. The method for generating 3D spatio-thermal models of building environments builds upon the authors' recently prototyped methods [(Ham and Golparvar-Fard 2013)] and (Golparvar-Fard and Ham [2014]). First, both thermal and digital imagery are collected by practitioners using a consumer-level IR camera. Then, the 3D geometrical point cloud of the building is automatically generated by using the collected digital images. Here, an image-based 3D reconstruction method is used, which consists of Graphic Processing Unit (GPU)-based Structure-from-Motion (SfM) and Multi-View Stereo (MVS) algorithms. The next step is 3D thermal modeling using the collected thermal images. Unlike digital images, thermal images are typically featureless in computer vision terms and have low spatial resolutions. The lack of distinct visual features impede the application of the same image-based 3D modeling algorithm used on digital images for 3D building geometrical modeling purposes. Instead, through a one-time process, the thermal camera is internally calibrated, and the relative pose of the thermal camera lens with respect to the digital camera lens on the IR thermal camera is estimated. Based on the obtained intrinsic and extrinsic thermal camera parameters, a 3D thermal point cloud model is generated using the dense reconstruction algorithm. The resulting 3D thermal and geometrical point cloud models are automatically superimposed, forming *3D spatio-thermal* point cloud models. Next, the outcome is converted to a 3D thermal mesh model by using *k-d* tree structure and the nearest neighborhood searching algorithm. In this process, a geometrical baseline mesh is formed by using the boundary points extracted from the building geometrical point cloud as the basis for the transformation. The nearest 3D thermal point with respect to each vertex is searched for in the baseline mesh, and then the associated thermal values are mapped to the vertices. Figure 2 illustrates 3D thermal mesh modeling of two different building environments (top and bottom: Case #1 and #2 respectively).

Here, different from previous works [(Ham and Golparvar-Fard 2013)] and (Golparvar-Fard and Ham [2014]), and to better deal with static occlusions in building environments, a threshold is used to identify those thermal points that have a certain distance from the geometrical baseline mesh. It is noted that the thermal points may belong to objects on or around the wall, not the building assembly itself. Thus, when searching for the nearest thermal points to each vertex in the baseline mesh, a geometrical threshold based on the registration error of the point cloud and the baseline mesh is applied. At a later stage, the points from non-relevant objects are discarded and are not used for calculations. The details will be discussed in Section (Considering static occlusions in 3D visualization of the as-is building conditions).

Figure 2. 3D thermal mesh modeling. **(a)**: Unordered digital and thermal imagery, **(b)**: Geometrical baseline mesh models, **(c)**: 3D thermal mesh models of interior surfaces jointly represented with building geometrical point cloud.

3.2 Measuring actual R-values at the level of 3D points

The environmental measurements in this paper build on the environmental assumptions presented in the section (Environmental Assumptions and Experimental Setups), which is consistent with recent works on non-destructive measurement of R (or U)-value using thermography (Albatici and Tonelli [2010]; Dall'O' et al. [2013]; Fokaides and Kalogirou [2011]; Ham and Golparvar-Fard [2014]; Madding [2008]). Under such environmental assumptions, the overall heat transfer rate (*dQ/dt*) through a building surface with the area of (*A*), a thermal resistance (*R*), and a temperature difference of *ΔT* between inside and outside can be described using the following equation:

$$\frac{dQ}{dt} = \frac{1}{R} \times A \times \Delta T \tag{1}$$

Here, the overall heat transfer (*Q*) in building environments can be described by the combination of thermal convection and radiation. First, the heat transfer caused by thermal convection between the interior surface of building environments and the indoor air is calculated using the following equation:

$$Q_{Convective} = \alpha_{convective} \times A \times \left| T_{inside,\ air} - T_{inside,wall} \right| \tag{2}$$

Where $\alpha_{convective}$ is the convective heat transfer coefficient which is a function of the airflow (e.g., laminar or turbulent) and the temperature deviations between building surfaces and indoor air. Typical convective heat transfer coefficients of indoor building environments are within the range of 2–8 W/m^2K. In this work, this value is adopted based on (Dall'O' et al. [2013]). As part of data collection, the inside and outside air temperatures, $T_{inside,\ air}$ and $T_{outside,\ air}$, are measured using a thermometer. Next, the heat transfer between the interior building surface and the surroundings caused by thermal radiation is calculated using the following equation:

$$Q_{Radiation} = \varepsilon \times \sigma \times A \times \left| T^4_{inside,\ wall} - T^4_{inside,\ reflected} \right| \tag{3}$$

Where ε is thermal emissivity and σ is Stefan-Boltzmann constant $(5.67 \times 10^{-8}\ W/m^2K^4)$. For thermal emissivity, the measurement in (Mikron Instrument Company [2012]; Singham [1962]) was adopted. To measure the reflected temperature ($T_{inside,\ reflected}$), a small crumpled aluminum foil was placed on the inspection areas prior to the thermographic inspection (Figure3). Since the crumpled aluminum foil has low emissivity and high reflectivity, the reflected temperature can be robustly measured from the crumpled foil (FLIR system [2010]). Prior works (Fokaides and Kalogirou

[2011]; Madding [2008]) experimentally demonstrated that measuring the difference between surface and reflected apparent temperatures is one of the most sensitive factors on reliable R-value measurement in a non-destructive form. Here, instead of extracting a single temperature from the crumpled foil or averaging temperatures of a few target points as used in (Fokaides and Kalogirou [2011]; Madding [2008]), the average temperature is calculated from all pixels that represent the crumpled foil. To do so, the RGB color values of these pixels from the thermal image (Figure 3c) were converted into the corresponding absolute temperature values based on a normalized temperature spectrum.

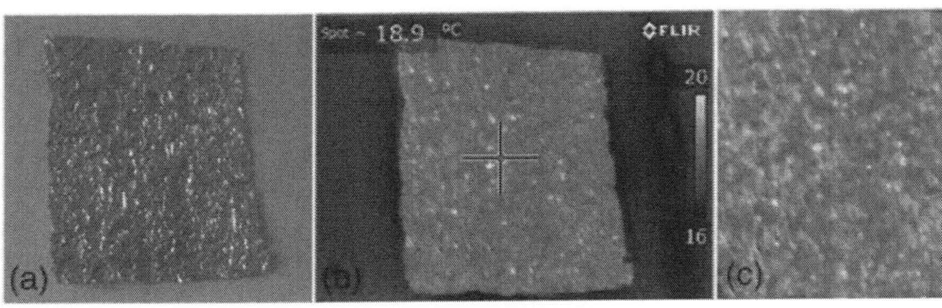

Figure 3. Measuring reflected temperature. **(a)**: crumpled aluminum foil located on the inspection surface, **(b)**: thermal image simultaneously captured from the same viewpoint, **(c)**: temperature readings on the crumpled aluminum foil surface.

It is assumed that interior surfaces in the given space under inspection have the same reflective temperatures, which is experimentally validated in (Fokaides and Kalogirou [2011]). Finally, by combining the Equations (1), (2), and (3), the thermal resistance can be described by using the following equation:

$$R = \frac{\left|T_{inside,\ air} - T_{outside,\ air}\right|}{\alpha_{con} \times \left|T_{inside,\ air} - T_{inside,wall}\right| + \varepsilon \times \sigma \times \left|T^4_{inside,\ wall} - T^4_{inside,\ reflected}\right|} \tag{4}$$

Within the resulting 3D spatio-thermal mesh models, $T_{inside,\ wall}$ can be queried from each 3D vertex. Here, different from prior works, this study focuses on measuring actual thermal resistance of building assemblies while taking possible static occlusions into account as discussed in the previous section.

3.3 Detecting building condensation problems in 3D

Building condensation is due to the fact that a given volume of air at a given temperature can only accommodate a certain amount of water. This is typically characterized by the relative humidity and the maximum absolute humidity. The absolute humidity is a measure of the amount of the water content in the air, which itself is a function of the air temperature. The relative humidity is a measure of the percentage of the actual water content in the air with respect to the maximum absolute humidity (i.e., the amount of maximum possible water content at a given air temperature). When the air contains more water content than the maximum possible amount at a given air temperature, the amount of the extra water vapor will condense on the surrounding surfaces. Hence, the dew point temperature (T_D) is a function of the relative humidity (f) and indoor temperature ($T_{inside,\,air}$) that can be described by the following equation (Wanielista et al. [1997]):

$$T_D = \left(\frac{f}{100}\right)^{1/8}\left(112 + 0.9T_{inside,air}\right) + 0.1T_{inside,air} - 112 \qquad (5)$$

The indoor temperature as well as the relative humidity are measured with the use of a thermo-hygrometer as part of the thermographic inspection. In this work, the building surface temperature ($T_{inside,\,wall}$) can be queried at the level of 3D vertices from the generated 3D spatio-thermal mesh models. Thus, the building surface areas that are likely to suffer from condensation problems are directly detected by comparing $T_{inside,\,wall}$ and T_D at the level of 3D points. Here, the building surface areas relevant to condensation issues can be calculated from the mesh model. For each triangle face consisting of the three thermal points that their thermal values are below T_D, the 3D coordinates of the vertices are queried, and then the area is calculated by using the cross product of two corresponding vectors that form the triangle face.

3.4 Considering static occlusions in 3D visualization of the as-is building conditions

For minimizing the impact of static occlusions when visualizing performance metrics and thermal problems obtained from the analysis of the 3D thermal model, non-relevant objects is segmented from the reconstructed 3D scene. The idea is that if the nearest thermal point in the 3D spatio-thermal model is too far from the vertices in the geometrical baseline mesh, it is most likely that the thermal points belong to objects around the wall but not the building geometry itself. Hence, a threshold is set based on the point cloud-to-mesh registration error. To do that, the registration accuracy of the 3D spatio-thermal point cloud with the geometrical baseline mesh in the site coordinate system is examined. Mapping thermal point clouds to a geometrical baseline mesh involves transforming the 3D spatio-thermal model into the real-world site coordinate system. To determine this transformation, the closed-form solution using unit quaternions is used (Horn [1987]). The transformation from the 3D spatio-thermal model into the real-world has 7 degrees of freedom (1 for scaling +3 for translation +3 for rotation), and thus requires a minimum of three matching control points between the 3D spatio-thermal model and the real-world coordinate system Ham and Golparvar-Fard ([2013a]). These control points, $P^i_{Thermal}$ and P^i_{Site}, can be extracted from the known dimensions in the building environments. This is solved by minimizing the sum of squared residual errors of the Euclidean transformation for (n) control points using the following:

$$\sum_1^n \|e^i\|^2 = \sum_1^n \|p^i_{Site} - sR\left(P^i_{Thermal}\right) - T\|^2 \tag{6}$$

Where, s, T, and $R\left(P^i_{Thermal}\right)$ indicate the scaling factor, translational offset, and rotation of 3D spatio-thermal models respectively. The resulting registration error (e) is then used as a threshold for detecting thermal points that are likely to represent static occlusions. Among all $T_{inside,\ wall}$ that can be queried from each 3D vertices within the 3D spatio-thermal mesh, those thermal points belonging to non-relevant objects (i.e., above the threshold (e)) are removed as outliers and are not considered in the calculation of the thermal resistance of building assemblies and detecting condensation problems. Once the relevant-to-building geometry thermal points are queried from 3D spatio-thermal mesh models, the actual R-values of the building assemblies under inspection are calculated by using

Equation (4), and the building areas associated with condensation issues are detected by using Equation (5).

Then, for 3D visualization, the measured R-values and the comparison of $T_{inside, wall}$ and T_D at the level of 3D points are converted into the corresponding RGB color value. To visualize the measured metrics and problems, the visualization platform provides several navigation modes such as a 3D free-flight navigation to efficiently visualize performance metrics and potential issues with the related building views. Similar to most engineering 3D viewers, when a camera is visited in the reconstructed 3D scene, the user can pan, tilt, and zoom-in & out for exploring the particular building areas in a mesh. For integrated visualization of the distribution of the actual R-value, the detected building areas with condensation issues, and the related geometrical and thermal characteristics, the proposed visualization platform leverages the following data structures:

1 $P_1^e, P_2^e, \ldots, P_n^e$: a set of relevant-to-building geometry 3D mesh vertices where $P_i^e = X_i^e, Y_i^e, Z_i^e, RGB_i^{R\text{-}value}, RGB_i^{Condensation}, RGB_i^{SurfaceTemperature}$ encapsulates the 3D location (X_i^e, Y_i^e, Z_i^e), the color value $(RGB_i^{R\text{-}value})$ of each vertex which corresponds to the measured R-value according to the given color spectrum, the color value $(RGB_i^{Condensation})$ of each vertex which is a metaphor based on traffic light colors to represent problematic and non-problematic areas using red and green colors respectively, and finally the color value $(RGB_i^{SurfaceTemperature})$ of each vertex associated with a temperature reading averaged from all thermal images that have observed the vertex

2 $P_1^g, P_2^g, \ldots, P_m^g$: a set of 3D points where $P_i^g = X_i^g, Y_i^g, Z_i^g, RGB_i^{Geometry}$ encapsulates the 3D location (X_i^g, Y_i^g, Z_i^g) and the color $(RGB_i^{Geometry})$ of each point in 3D building geometrical point cloud which is the average from the colors in all digital images that observed the point

3 $C_1^T, C_2^T, \ldots, C_l^T$: a set of thermal cameras representing the thermal camera projection matrix based on the relative pose with respect to the built-in digital camera

4 $C_1^D, C_2^D, \ldots, C_l^D$: a set of built-in digital cameras representing the digital camera projection matrix for each camera registered in a reconstructed 3D scene

5 A set of vertices that form each face

6 The number of vertices and faces in a mesh

7 The number of feature points and cameras that observe the points

8 Mapping relationship between points and cameras that observe the points

4. RESULTS AND DISCUSSIONS

The first five rows in Table 2 present the 3D thermal modeling metrics and the results from the case studies. In both cases, the computational time is benchmarked on a regular engineering workstation (an Intel i7 960 core with 24GBs RAM and NVIDIA GeForce GTX 400), and reasonably detailed 3D mesh models are generated in each case. Table 2 also shows the environmental measurements that were conducted during the experiments. Here, rather than concluding a single value of the actual thermal resistance for a building assembly in a deterministic way, Table 2 presents the range that the R-value for inspected building assemblies is most likely to be included.

Figures 4, 5, 6, and 7 illustrate the distribution of the surface temperatures and the measured R-values at 3D points for wall assemblies and windows in case #1 and #2 respectively. Compared to prior works (Albatici and Tonelli [2010]; Dall'O' et al. [2013]; Fokaides and Kalogirou [2011]; Madding [2008]) which leverage a single measurement of actual thermal resistance for each building assembly using a single or a few temperature data, the proposed method can provide more detailed representation of the heat transfer conditions for building assemblies at the level of 3D points. Figures 8 and 9 show examples of visualizing the distribution of the R-values along with the corresponding building geometry in 3D. The 3D visualization of thermal resistances using different color gradients can provide an opportunity for building practitioners to easily and quickly understand the distribution of the actual thermal resistance in building environments. Because static occlusions are taken into account, the proposed method has potential for improving the accuracy of measurements. In both cases, by using the equation (6), the registration errors between 3D thermal point cloud and the geometrical baseline mesh are within a reasonable practical range (Table 2). Based on the calculated registration errors, it is observed that 8.8% (case #1) and 43.8% (case #2) of the thermal points were related to non-relevant objects around wall (e.g., Figures 8 and 9(b) and (c)). Thus, those thermal points are not considered when calculating the thermal resistances in 3D. Although it is attempted to minimize the impact of such static occlusions, false negatives are still possible due to incomplete segmentation. Examples are shown as oval-shaped areas in the Figures 8 and 9(e), which causes inaccurate thermographic inspection for those areas. Here, the proposed method can also be used for the purpose of commissioning new buildings in addition to retrofit purposes during the operational phase. Unlike occupied buildings, there is minimal occlusion during commissioning inspection. Here, the proposed method can inspect if the new building is operating as expected by the design.

Beyond providing surface temperature data, the proposed method can also improve the quality of interpreting building thermography by detecting and localizing moisture problems on building envelopes and visualizing the outcome in 3D along with the corresponding geometrical characteristics. Figures 4, 5, 6, and 7 illustrate the dew point temperature (dotted line) and the distribution of surface temperatures for both case studies. These figures demonstrate noticeable deviations between these two temperatures and further highlight the prominence of the point-based measurements. Compared to the built-in option available in current thermal cameras (Hoff [2011]) for detecting condensation problems, the proposed method has potential to improve the current practices by estimating the size of the building areas associated with condensation issues. In Figure 10, building surface areas that are likely to suffer from condensation issues are visualized in 3D ((a) in Case #1 $\cong 0.0046m^2$, (b) and (c) in Case #2$\cong 0.0159m^2$). These results are consistent with inspection observations conducted at the time of data collection. When thermal images were captured, the surface temperatures of these areas were lower than the dew point. As shown in both cases, the condensation issues typically starts from around windowsills. To minimize the impact of these detected condensation problems, homeowners can (1) reduce the dew point temperature by decreasing the relative humidity or the indoor air temperature; (2) replace or add new insulation layers to increase the surface temperature on the areas where insulation layers have been missed or degraded; and finally (3) adjust the ventilation fan or better benefit from natural ventilation to increase the wind velocity around the problematic areas.

Table 2. Experimental results of 3D thermal modeling and environmental measurements

Item	Metrics	Case #1	Case #2
3D thermal modeling	# of 2D Thermal Images	446	429
	# of 3D Thermal Points	3,042,350	2,838,478
	# of Vertices in Baseline Mesh Model (wall)	18,964	13,771
	# of Vertices excluding static occlusions in Mesh Model by using Threshold (wall)	17,297	7,740
	Computational Time for 3D Thermal Modeling	2hr 30min	2hr 21min
Environmental measurements	Outside air temperature (°C)	5	3
	Indoor air temperature (°C)	28	24
	Dew point temperature (°C)	13.15	11.64
	Relative humidity (%)	40	46
	Error on registration (mm)	8.69	9.30
Calculated metrics in 3D	Average (μ) of R-values of Wall (m^2K/W)	0.753	0.833
	Standard Deviation (σ) of R-values of Wall	0.10207	0.14463
	Range of actual R-values by considering 95% confidence ($\mu-1.96\sigma/\sqrt{n}$, $\mu+1.96\sigma/\sqrt{n}$)	(0.75168, 0.75472)	(0.82986, 0.83631)
	Average (μ) of R-values of Window (m^2K/W)	0.260	0.274
	Standard Deviation (σ) of R-values of Window	0.00173	0.00211
	Range of actual R-values by considering 95% confidence ($\mu-1.96\sigma/\sqrt{n}$, $\mu+1.96\sigma/\sqrt{n}$)	(0.26011, 0.26027)	(0.2737, 0.27389)

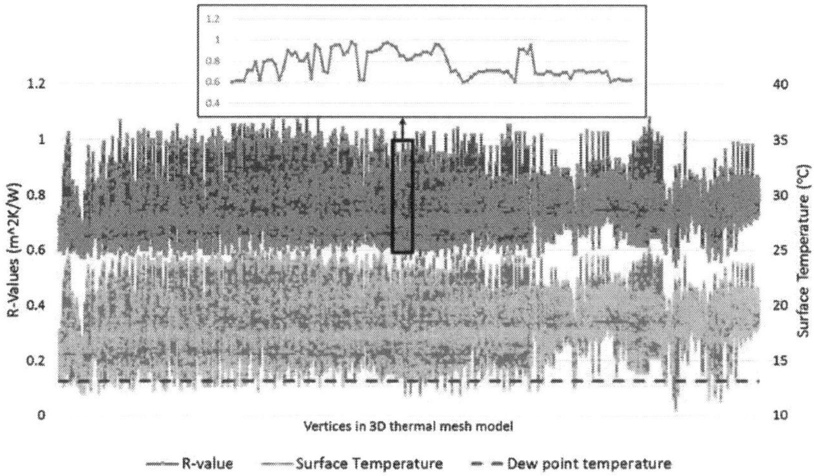

Figure 4. Point-level temperature and actual R-values for the wood frame walls in case study #1.

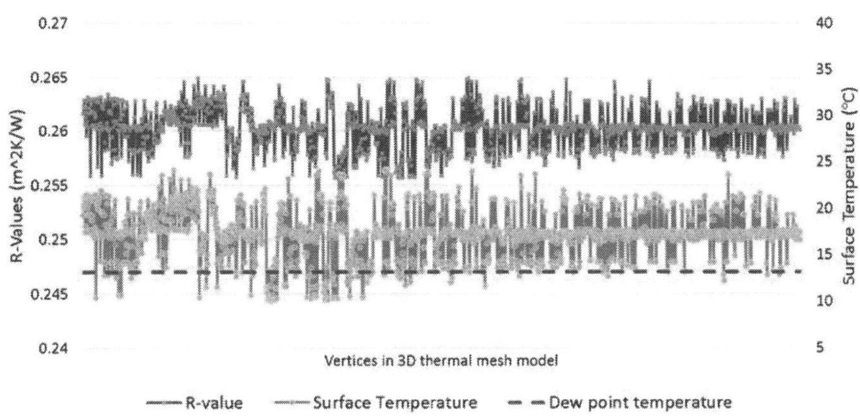

Figure 5. Point-le el temperature and actual R-values for the single-glazed clear window in case study #1.

The timing of an inspection would impact the capability of detecting condensation issues. To continuously monitor changes in the environmental conditions and study their impacts on condensation problems, one can place environmental sensors on building interior surfaces and continuously monitor the relative humidity and also the surface and inside air temperatures. However, in practice, only a few environmental sensors could be placed per element. As such, considering the variations of thermal performance over surface areas, it is difficult to ensure if the surface condition measured by a few sensors is representative of the entirety of the building assemblies. Here, the proposed method has potential to provide a solution for selecting the proper locations of installing a limited number of environmental sensors for detecting condensation problems. As a complementary step, after detecting the condensation problems from the proposed one-time thermography-based method, the environmental sensors can be placed at the detected areas for continuous monitoring. This will allow proper positioning of a limited number of sensors and improves the efficiency of monitoring.

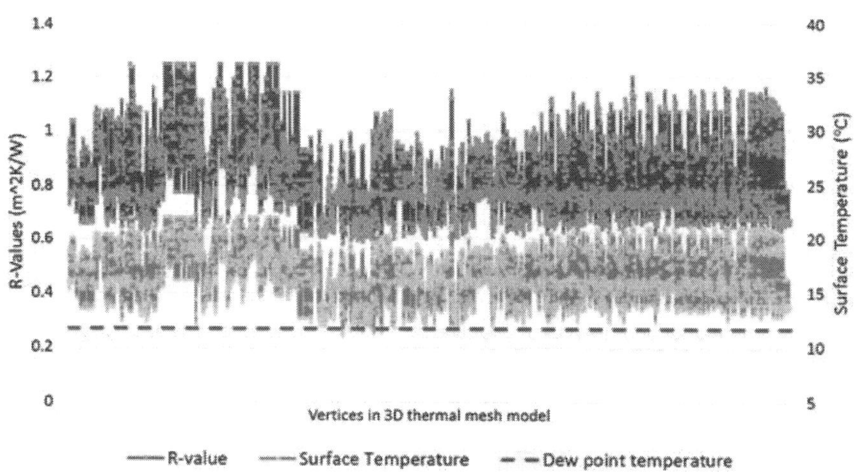

Figure 6. Point-level temperature and the actual R-values for concrete masonry walls in case study #2.

4.1 Domain expert evaluation and discussions

To discuss the perceived benefits of the proposed method, face-to-face and phone interviews were conducted with eleven actual domain experts. The subjects had 2 to 10 years of practical experiences in energy auditing of existing buildings. They were first asked about the current best practices and bottlenecks in thermographic inspections for energy diagnostics in existing buildings. All subjects responded that today, thermography captured from existing buildings is considered the most effective tool for recording and communicating the as-is energy performance conditions to building owners. However, thermography is primarily used for qualitative documentations in support of leveraging the Home Energy Rating System (HERS) for reporting purposes at a later stage. Most responders used the word "qualitative" to describe the current process. For example, they noted that "Audits can take a long time, up to several hours for large residential houses. IR imaging is done piece-wise through a building. It takes detailed documentation to know where each image was taken and what it is supposed to be showing. This can be confusing, and is often detrimental to the client's understanding, even with a good service provider explaining discoveries.", "Sometimes it is hard to tell where the thermal image is coming from in the building. Often, our audits last about 2–4 hours. In this time, it is hard to note which image was taken where within the building."

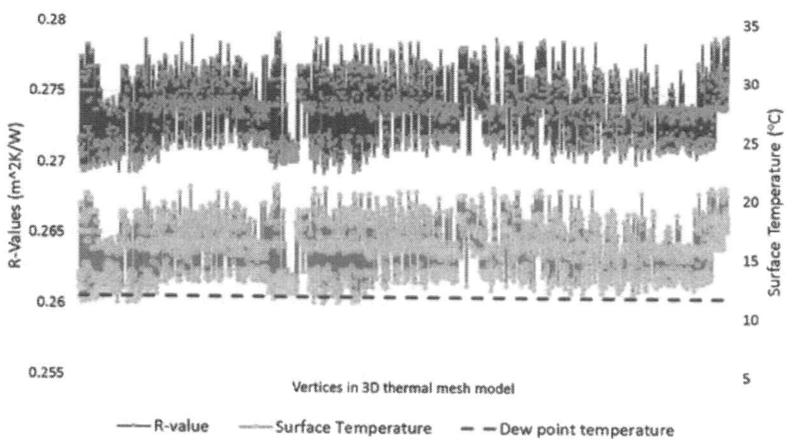

Figure 7. Point-level temperature and the actual R-values for single-glazed clear windows in case study #2.

Figure 11 illustrates the summary of the responses from the Likert questionnaires on the usefulness of each feature (1 = poor, 5 = excellent). As shown, participants generally felt that each features in the proposed work would be helpful for reducing the bottlenecks in current practices on thermography-based building condition assessment, e.g., "This is a very interesting step in the direction of recalling and identifying locations of leakage. This visualization tool will dramatically increase the potential for accurate assessments of issues.", "This visualization tool can be very efficient in detecting leaks in building insulation.", "In the summary, it would be useful to see an entire room in 3D to capture the proposed visualization tool's differentiating usefulness that cannot be done with 2D imaging."

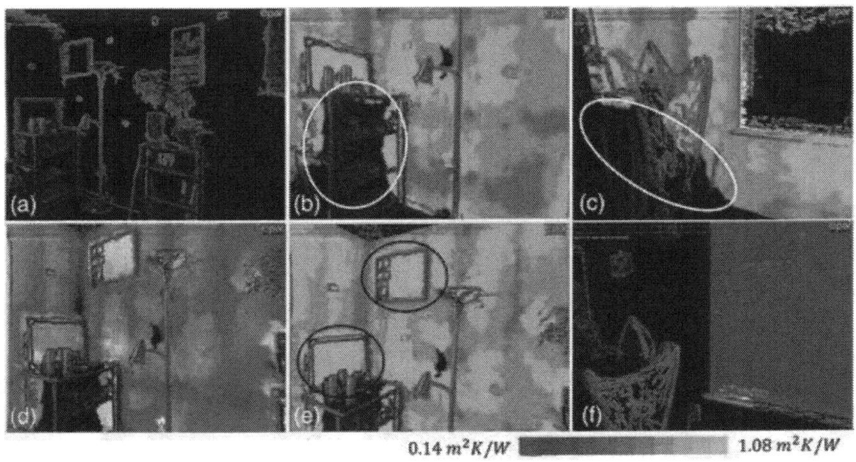

Figure 8. 3D visualization along with the building geometry for case #1. The color spectrums shows variation between 0.14 to 1.08 m2K/W.**(a)**: 3D reconstructed built environments, **(b)** and **(c)**: thermal resistances of exterior walls with True Positive detection of occlusion,**(d)**: 3D thermal distribution with True Positive detection of occlusion, **(e)**: thermal resistances of exterior walls with False Negative detection of occlusion, and **(f)**: thermal resistances of a window component.

The survey results show the potential of the proposed 3D visualization method for assessing the as-is building conditions as an intuitive communication tool. Since the main purpose of the proposed method is to provide guidance for making efficient and effective energy retrofit decisions by diagnosing and visualizing the as-is building conditions, the proposed method only needs to be conducted once per regular building diagnostic process (e.g., annual building inspection) and does not need to be performed more frequently. Hence, it does not add significant burden to the building practitioners, and allows the proposed method to be repeatable in parallel with thermographic inspection. The proposed method can inspect very small surfaces with potential performance problems (in fact, at the vertex-level of a mesh). This level of granularity, for example, would be helpful when installing blown-in loose-fill insulation is considered as one of the retrofit alternatives to improve the thermal resistance of the defected areas. However, if the detected areas are too small, retrofitting such problems is not likely to be practically feasible due to technical challenges and long payback period. Thus, practical retrofit decisions need to be taken up to a certain minimum size of retroffitable surface. The proposed method chooses not to dictate the minimum size of the retrofittable building area. Rather, by presenting the measurements at the point-level, the proposed method lets the building practitioners set their desired value prior to assessing different energy efficiency retrofit alternatives.

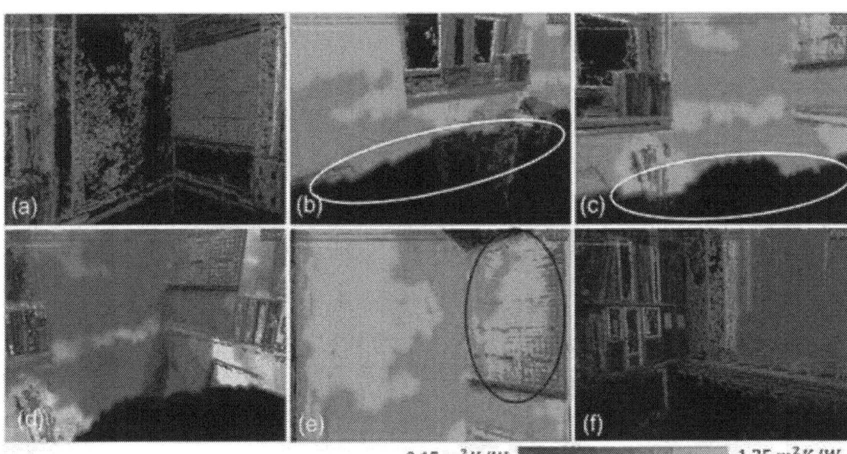

Figure 9. 3D visualization along with the building geometry for case #2. The color spectrums shows variation between 0.15 to 1.25 m2K/W.**(a)**: 3D reconstructed built environments, **(b)** and **(c)**: thermal resistances of exterior walls with True Positive detection of occlusion,**(d)**: 3D thermal distribution with True Positive detection of occlusion, **(e)**: thermal resistances of exterior walls with False Negative detection of occlusion, and **(f)**: thermal resistances of a window component.

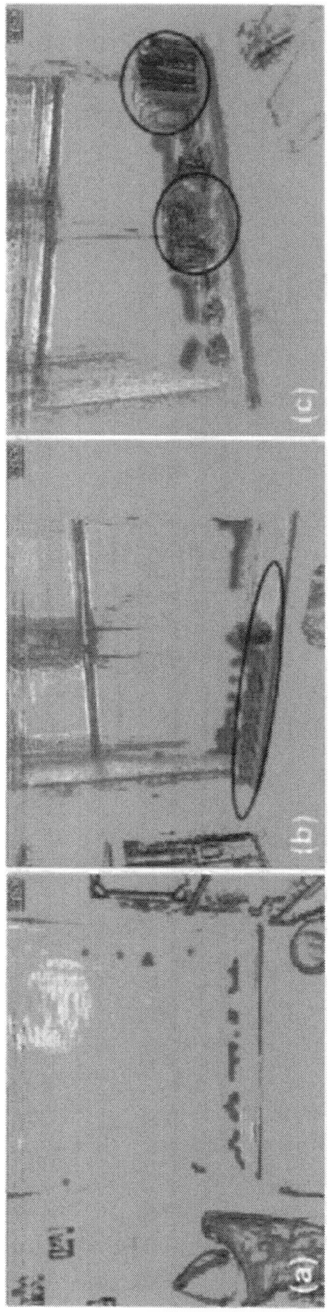

Figure 10. 3D visualization of the building areas identified with possible condensation issues (the color-coding is based on the metaphor of traffic light colors) (a): case #1, (b) and (c): case #2.

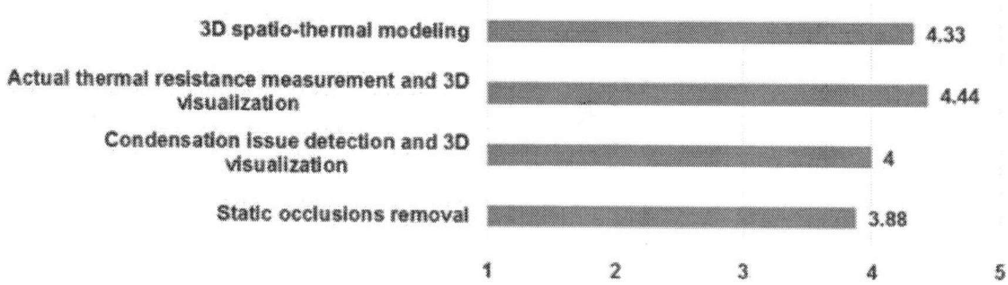

Figure 11. Summarized responses (averaged over all subjects) from the Likert scale questionnaires about the usefulness of each feature (1 = poor, 5 = excellent).

Despite such perceived and potential benefits, several participants noted that there is still room for improvement to alleviate bottlenecks in current practices, which are remaining open:

1 *Difficulty in in-situ measurement of as-is thermal properties for building assemblies*: Few participants had concerns about the difficulty in accurate measurement of thermal properties in building environments. Here, following prior works (Albatici and Tonelli [2010]; Dall'O' et al. [2013]; Fokaides and Kalogirou [2011]; Madding [2008]), the deviation between the measurements and the notional values was examined. It is noted that wall assembles are not the best choice for such experiments since wall assemblies typically have multiple layers and are likely to suffer from non-uniform and time-variant degradations. Hence, in this case, using the notional R-values as the ground truth for the as-is thermal resistances may not be accurate. In contrary to walls, since windows are mostly standardized and their thermal properties are more likely to be consistent over time, their actual thermal resistance is less likely to deviate from the notional values derived from the as-built documents. Thus, window components were selected for experiments. It is not a trivial task to accurately estimate the thermal properties of the old building assemblies such as the case studies of this paper. Here, the estimation is based on (ColoradoENERGY [2013]; The Engineering Toolbox [2013]) which provide recommendations for such R-values and (European Standard (EN) [1997]). As a result for single-glazed clear windows, the averaged deviations were 12.81% (= (0.301-0.260)/0.301) for case #1 and 10.37% (= (0.301-0.274)/0.301) for case #2. Following the environmental assumptions presented in (Albatici and Tonelli [2010]; Dall'O' et al. [2013]; Fokaides and Kalogirou

[2011]; Madding [2008]), the findings in the experiments are consistent with what was reported in those literatures (Section Thermal resistance (R-value) of building assemblies). Here, systematic errors on measuring the input variables for Equation (5) need to be considered. If the thermographic inspection is conducted in outdoor environments, the measurement accuracy would be affected by various external factors such as wind, sun radiation, and shadows. Since the temperatures are measured from inside surfaces and crumpled aluminum foils located in indoor environments, the accuracy of measuring those input variables is typically the same as it is claimed in the technical specification of the thermal camera. As can be seen in the Table 1, the measurement accuracy of the consumer-level thermal camera used is 2°C or ±2% of reading. Here, another prevalent challenge for accurate measurement in thermal resistances in a non-destructive manner is that the measurement needs to be conducted under steady-state conditions as discussed in Section (Environmental Assumptions and Experimental Setups). Despite significant efforts to form a steady-state heat transfer condition during thermographic inspection (e.g., the timing of data collection and experimental setups prior to image data collection as described in Section(Environmental Assumptions and Experimental Setups), in practice, it is not trivial to maintain a perfect steady-state condition during thermogaphic inspection. More research needs to be conducted on how a steady state condition of heat transfer can be continued during thermographic inspection and how the measurement errors caused by non-steady-state conditions can be accounted for and minimized for more accurate analysis;

2 *Continuous measurement to form time-series of thermal resistances*: Besides the measurement accuracy issues, a few responders described `4D visualization' as a desired feature, e.g. "Maybe a 4D visualization done with the presence of all the objects that exist in the building.", "A time series IR webcam of a building over the course of a year." In prior work on measuring actual U-values (Fokaides and Kalogirou [2011]), the results were validated based on different sources of information: instantaneous surface temperature data samples of 10 measurements obtained from a thermal camera and continuous time-series heat flux data collected from HFMs over a period of 168 hours. According to the experimental results, it is shown that the outcomes of the two methods were very close to each other in terms of their accuracies. However, if surface temperature data can be collected continuously, the actual heat resistances can be modeled and explored in 4D (3D + time) as opposed to static information. Visualizing dynamic energy

performance metrics in form of a 3D animation can significantly improve the understanding of time-varying heat transfer phenomena. This also enables to provide an opportunity to analyze the impacts of building envelope retrofit decisions. For example, energy auditors can conduct a before/after analysis on thermal resistances when additional insulation layers are being installed. Such analyses would support implementation of periodic building maintenance programs. More research on 4D visualization of energy performance metrics need to be conducted; and finally

3 *Minimizing false negative detection for static occlusions*: One survey participant noted that "Accounting for furniture and other obstructions seems like a challenging task". The proposed method of setting a geometrical distance threshold has potential to ignore those building areas with static occlusions. Nevertheless, as discussed in the previous section, false negatives can still occur due to incomplete segmentations. In the experiments, those happen around thin objects on walls such as frames (e.g., Figures 8 and 9(e)) or tiny objects around windows (Figure 10(b) and (c)). This is because the distance between thermal points representing such objects in 3D point clouds and the nearest vertex in a geometrical baseline mesh is most likely to be within the threshold based on the registration error. As a result, the thermal resistance of the occluded building assemblies was not accurately calculated. More research needs to be conducted to better segment non-relevant objects from the reconstructed 3D scenes.

5. CONCLUSIONS

Effective building diagnostic tools should represent abnormal building conditions in an intuitive format to allow energy auditors better communicate the as-is building conditions to the homeowners. Toward this goal, this paper presents a method for visualizing the as-is building condition assessment at the level of 3D points. The outcome of the analyses are visualized within a 3D environment, and static occlusions are accounted for. In this 3D environment, practitioners can navigate and interactively query information about the building geometry, surface temperature data, thermal resistances, and analyze condensation problems. The proposed method was tested in a residential building and an instructional facility, and the following domain expert evaluation was conducted. By converting surface temperature data sensed from a consumer-level single thermal camera into 3D visualization of the actual R-value distributions and the location information of possible condensation problems, building practitioners can better understand the as-is building conditions. Future works include addressing the following challenges in (1) forming and maintaining a steady state condition of heat transfer; (2) studying 4D visualization of energy performance metrics based on time-series of building surface temperatures; (3) minimizing false negative segmentations for static occlusions; and (4) Varying the level of visualization for both global and local inspection of as-is energy performance based on different Levels of Detail (LOD). These are currently being explored as part of ongoing research.

AUTHORS' CONTRIBUTIONS

YH designed and implemented the proposed 3D visualization of as-is building conditions, carried out the thermographic experiments and domain expert evaluations, and drafted the manuscript. MGF supervised the entire processes of this study, co-designed the 3D visualization tool, and edited the manuscript. All authors read and approved the final manuscript.

REFERENCES

1. Aelenei D, Henriques FMA: Analysis of the condensation risk on exterior surface of building envelopes. *Energy and Buildings* 2008,40 (10):1866–1871.

2. Albatici R, Tonelli AM: Infrared thermovision technique for the assessment of thermal transmittance value of opaque building elements on site. *Energy and Buildings* 2010, 42(11):2177–2183.

3. Bellia L, Minichiello F: A simple evaluator of building envelope moisture condensation according to an European Standard.*Building and Environment* 2003, 38(3):457–468.

4. Borrmann D, Elseberg J, Nüchter A: Thermal 3D Mapping of Building Façades. In *Intelligent Autonomous Systems 12*. Edited by: Lee S, Cho H,

Yoon K-J, Lee J. Springer Berlin Heidelberg, Berlin, Germany; 2013:173–182.

5. ColoradoENERGY. (2013). R-Value Table: Insulation Values For Selected Materials. http://www.coloradoenergy.org/procorner/stuff/r-values.htm.

6. Dall'O' G, Sarto L, Panza A: Infrared screening of residential buildings for energy audit purposes: results of a field test. *Energies* 2013, 6(8):3859–3878.

7. European Standard (EN). (1997). Glass in Building - Determination of Thermal Transmittance (U-value) - Calculation Method (EN 673).

8. FLIR system. (2010). Thermographhic Measurement Techniques (FLIR Reporter Professional, Professional Edition, pp. 133–137).

9. Fokaides PA, Kalogirou SA: Application of infrared thermography for the determination of the overall heat transfer coefficient (U-Value) in building envelopes. *Applied Energy* 2011, 88(12):4358–4365.

10. Golparvar-Fard M, Ham Y: Automated diagnostics and visualization of potential energy performance problems in existing buildings using EPAR -Energy Performance Augmented Reality-models. *Journal of Computing in Civil Engineering* 2014, 28(1):17–29.

11. Ham Y, Golparvar-Fard M: An automated vision-based method for rapid 3D energy performance modeling of existing buildings using thermal and digital imagery. *Advanced Engineering Informatics* 2013, 27(3): 395–409.

12. Ham Y, Golparvar-Fard M: EPAR: Energy Performance Augmented Reality models for identification of building energy performance deviations between actual measurements and simulation results. *Energy and Buildings* 2013, 63: 15–28.

13. Ham, Y, & Golparvar-Fard, M. (2014). Three-dimensional thermography-based method for cost-benefit analysis of energy efficiency building envelope retrofits. Journal of Computing in Civil Engineering.,

14. Hoff R: *Water condensation – a Temperature Related Phenomenon*. 2011. http://www.moistureview.com/blog-034.

15. Horn B: Closed-form solution of absolute orientation using unit quaternions. *Journal of the Optical Society* 1987, 4(4):629–642.

16. International Organization for Standardization (ISO). (2007). Building Components and Building Elements-Thermal Resistance and Thermal Transmittance-Calculation Method (ISO 6946).

17. Lagüela S, Díaz-Vilariño L, Martínez J, Armesto J: Automatic thermographic and RGB texture of as-built BIM for energy rehabilitation purposes. *Automation in Construction* 2013, 31: 230–240.

18. Madding R: *Finding R-Values of Stud Frame Constructed Houses with IR Thermography*. Proc, InfraMation; 2008.

19. Mikron Instrument Company. (2012). Table of Emissivity of Various Surface.

20. Roth K, Westphalen D, Feng M, Llana P, Quartararo L: *The Energy Impact of Commercial Building Controls and Performance Diagnostics: Market Characterization, Energy Impact of Building Faults and Energy Savings Potential*. TIAX LCC Technical Report Cambridge, MA; 2005.

21. Singham JR: Tables of emissivity of surfaces. *International Journal of Heat and Mass Transfer* 1962, 5(1–2):67–76.

22. The Engineering Toolbox. (2013). Heat Loss through Building Elements due to Transmission. http://www.engineeringtoolbox.com/heat-loss-transmissiond_748.html.

23. U.S. DOE. (2010). 2010 U.S. DOE Buildings Energy Databook. Washington D.C.: U.S. Department of Energy.

24. Wang C, Cho Y, Gai M: As-Is 3D Thermal modeling for existing building envelopes using a Hybrid LIDAR system. *Journal of Computing in Civil Engineering* 2013, 27(6):645–656.

25. Wanielista M, Kersten R, Ron E: *Hydrology Water Quantity and Quality Control*. John Wiley & Sons, Hoboken, NJ; 1997.

Chapter 3

Planned energy-efficient retrofitting of a residential building in Italy

Enzo Zanchini Claudia Naldi, Stefano Lazzari and Gian Luca Morini

ABSTRACT

The planned energy-efficient retrofitting of a residential building in Bologna, North-Center Italy is presented. The building is a detached house with an unheated basement, three floors with 2 apartments each, and an unheated attic. The total heated floor area is 281.9 m^2. The external wall is made of solid brick masonry and most windows are single glazed; no thermal insulation is present. Space heating is supplied by a gas boiler and radiators in the rooms. DHW is supplied by single-apartment electric boilers in 5 apartments and by a gas boiler in one apartment. Lighting is obtained by incandescent lamps.

The proposed retrofitting includes: external thermal insulation of the vertical walls by calcium silicate hydrates and loft insulation by mineral wool; replacement of windows; installation of a multifunction air-to-water heat pump for heating, cooling and DHW; replacement of the radiators by new heat exchangers; LED lighting; installation of PV panels. The building has been simulated by TRNSYS 17, and the heat pump has been simulated by own MATLAB codes. The retrofitting will reduce the total annual use of primary energy (excluding appliances) from 332.5 to 44.8 kWh/m^2, and will yield an important improvement of thermal comfort.

KEYWORDS

Energy retrofitting, Residential building, Multifunction heat pump, PV panels, Hourly simulation

1. INTRODUCTION

The economic growth of the 20th century has been based on a progressive increase of the world annual use of fossil fuels. The world annual use of primary energy is still increasing, and fossil fuels represent even now the most important source of primary energy, as shown in Fig. 1, which illustrates the world annual use of primary energy by source from 1980 to 2011, according to EIA (US Energy Information Administration) (http://www.eia.gov/cfapps/ipdbproject/IEDIndex3.cfm#). In 2011, 86 % of the world primary energy use is due to oil, carbon and gas. The fossil-fuel-based development has caused two important problems: the reserves of oil and natural gas are decreasing and the emission of carbon dioxide and of other greenhouse gases is causing a climate change (https://www.ipcc.ch/report/ar5/wg1/). As a consequence, all the industrialized and developing countries and, most of all, the European Union, are struggling to shift the economic growth towards a sustainable development, based on two main pillars: the increase of energy efficiency and the use of renewable energy sources. The energy policy of the European Union already obtained some success: the annual use of primary energy of the Union is slightly decreasing from 2006, as shown in Fig. 2. The figure illustrates the use of primary energy of the European Union by sector from 1990 to 2012, according to Eurostat (European Commission portal for statistics) (http://ec.europa.eu/eurostat/data/database); it reveals that the fractions of energy use in the residential sector and in the service sector are quite relevant. The fractions of primary-energy use in sectors for 2012 are better evidenced in Fig. 3, where it is shown that the sum of the fractions which refer to the residential sector and to the service sector, i.e., the total fraction due mainly to building operation, is 39.7 %. As a consequence, an important step towards the reduction of the use of fossil fuels in Europe would be the enhancement of the energy efficiency of buildings.

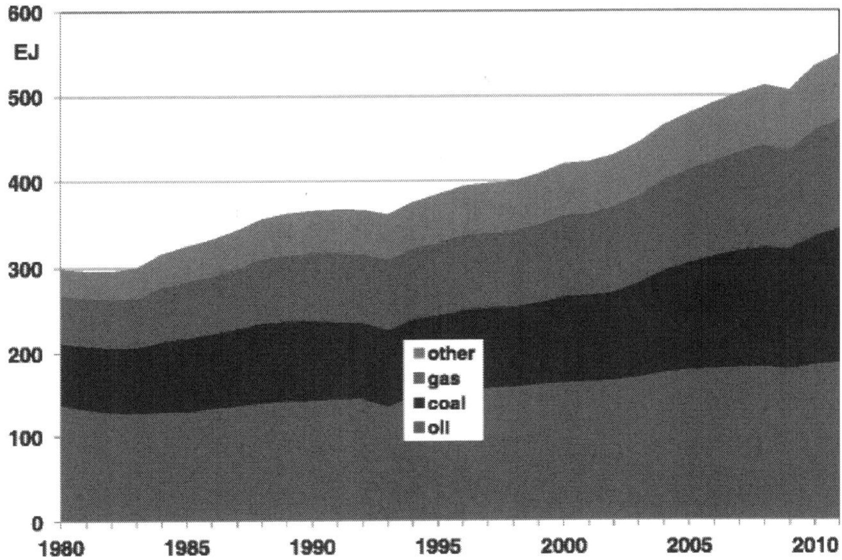

Figure. 1. World annual use of primary energy by source from 1980 to 2011, according to EIA

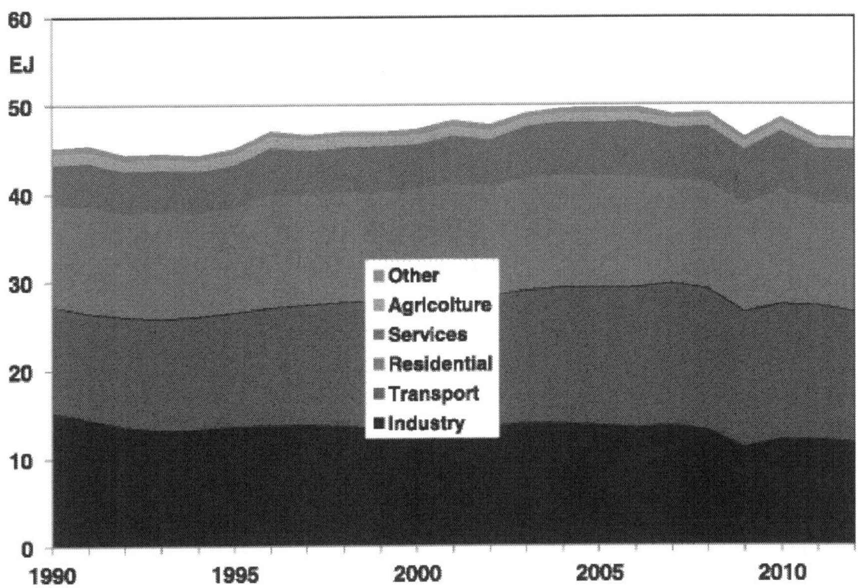

Figure. 2. Annual use of primary energy by sector in Europe, from 1990 to 2012, according to Eurostat

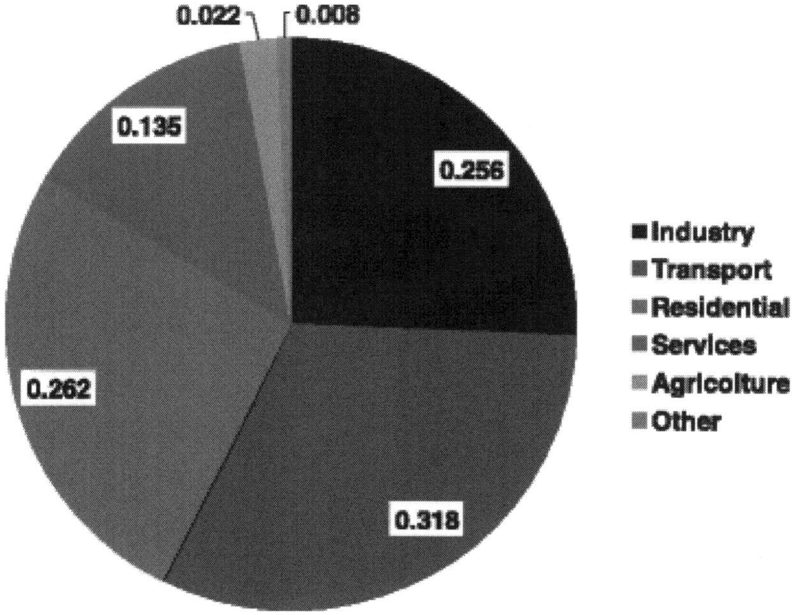

Figure. 3. Fractions of primary energy use in main sectors in Europe, in 2012, according to Eurostat

According to an official document of the European Commission (http://www.ectp.org/cws/params/ectp/download_files/36D2981v1_Eeb_cPPP_Roadmap_under.pdf), buildings use 40 % of the total EU energy consumption and generate 36 % of greenhouse gases in Europe. The construction sector is expected to reduce its CO_2 emissions by at least 80 % and its energy consumption by as much as 50 % by 2050. Since the replacement rate of the existing stock is very small (1–2 % per year), the energy retrofitting of existing buildings should play an important role to reach this goal.

A wide research activity on the techniques for energy retrofitting of buildings has been performed in the last decades. Some studies have concerned the optimization of thermal insulation (Ucar and Balo 2009; Yu et al. 2009; Daouas 2011; Cuce et al.2014), others the improvement of the plant efficiency (Bizzarri and Morini 2006; Zhao et al. 2008, 2009; Chan et al. 2010; Terlizzese and Zanchini 2011; Brignoli et al. 2013; Liu et al. 2013; Naldi et al. 2014), others the energy-saving potential (Balaras et al. 2000; Amstalden et al. 2007; Ascione et al. 2011; Chidiac et al. 2011a, b; Dall'O' and Sarto 2013; Shahrokni et al. 2014), others again the effects of solar chimneys, permeable coverings, cool roofs and green roofs (Afonso and Oliveira 2000; Orosa and Oliveira 2009; Pisello et al. 2013; Santamouris et al. 2011; Santamouris 2014).

Ucar and Balo (Ucar and Balo 2009) determined the optimum insulation thickness of external walls for four cities with different climatic conditions in Turkey, with reference to Foamboard, extruded polystyrene and fiberglass as insulation materials. Yu et al. (Yu et al. 2009) studied the optimization of the insulation thicknesses of expanded polystyrene, extruded polystyrene, foamed polyurethane, perlite, and foamed polyvinyl chloride, for a typical residential wall in China. Daouas (Daouas 2011) evaluated, by an analytical method based on Complex Finite Fourier Transform, the optimum insulation thickness, the energy saving and the payback period for a typical wall structure in Tunisia, in the presence of both cooling and heating loads. Cuce et al. (Cuce et al.2014) analyzed the optimum thermal insulation thickness of aerogel and its environmental impacts for the climatic conditions of Nottingham, UK.

Bizzarri and Morini (Bizzarri and Morini 2006) evaluated the reduction in pollutant emissions obtainable by installing new hybrid plats in hospitals. Several hybrid schemes were investigated and compared: phosphoric acid fuel cells, solar thermal systems and PV solar systems. Zhao et al. (Zhao et al. 2008) designed and numerically studied a novel dew point air conditioning system, and Zhao et al. (Zhao et al. 2009) investigated the feasibility of this system in several China regions. Chan et al. (Chan et al. 2010) pointed out advantages and limitations of passive solar heating and cooling technologies and suggested research guidelines to improve the economic feasibility of these techniques. Terlizzese and Zanchini (Terlizzese and Zanchini 2011) studied, through an economic and an exergy analysis, the feasibility of alternative plants for a zero carbon building complex in Italy. Brignoli, Cecchinato, and Zilio (Brignoli et al. 2013) performed an experimental investigation on air-to-water heat pumps, and compared a multiport aluminum flat-tube heat exchanger to a round-tube finned one. Liu et al. (Liu et al. 2013) studied a new kind of heat pump system, which utilizes gray water as heat source and sink for heating and cooling of residential buildings. Naldi et al. (Naldi et al. 2014) developed a MATLAB code for the choice of the optimal balance-point temperature of air-to-water heat pumps for heating.

Balaras (Balaras et al. 2000) audited 8 apartment buildings, located in three climatic zones of Greece, and showed that a considerable energy saving in heating, air conditioning, DHW production and lighting can be obtained by proper retrofit actions. Amstalden et al. (Amstalden et al. 2007) investigated the profitability of energy-efficient retrofit investments in the Swiss residential building sector from the house owner's perspective. Ascione, de Rossi and Vanoli (Ascione et al. 2011), through a dynamic simulation code experimentally calibrated, analyzed the effectiveness of several energy retrofit solutions for a historical building in Italy. Chidiac et.al. developed a methodology, based on the simulation code EnergyPlus, to determine the energy saving potential in the Canadian office building stock (Chidiac et al. 2011a); they also analyzed the effectiveness of single and multiple energy retrofit measures on the energy consumption of office buildings (Chidiac et al. 2011b). Dall'O' and Sarto (Dall'O' and Sarto 2013) investigated the technical end economic potential for increasing energy efficiency of 49 school building complexes in Lombardia (Northern Italy). Shahrokni et al. (Shahrokni et al. 2014) evaluated the energy efficiency potential in Stockholm, and showed that the retrofitting of the building stock to current building codes would reduce heating energy use in the city by one third.

Afonso and Oliveira (Afonso and Oliveira 2000) studied numerically and experimentally the increase in ventilation rate obtainable by solar chimneys. Orosa and Oliveira (Orosa and Oliveira 2009) investigated the effects of indoor permeable coverings on thermal comfort and energy saving in several office buildings in Spain. Pisello, Santamouris and Cotana (Pisello et al. 2013) analyzed the coupled passive–active effect produced by a cool roof on industrial-office building located in Rome. Santamouris, Synnefa and Karlessi (Santamouris et al. 2011) and Santamouris (Santamouris 2014) presented review papers on the positive effects obtainable by advanced cool materials and green roofs.

The EU-funded Project HERB (Holistic Energy-efficient Retrofitting of residential Buildings) started in October 2012, and aims to develop innovative technologies for the energy retrofitting of buildings and to perform demonstrations of holistic energy-efficient retrofitting of residential buildings in seven European Countries: United Kingdom, Italy, Portugal, Greece, Spain, Switzerland and Netherlands. In Italy, the demonstration concerns a residential building located in Bologna, North-Center Italy, a detached social house with 6 apartments owned by the Municipality of the city, with a total heated floor area of 281.9 m^2. The external wall is made of solid bricks and is uninsulated; most windows are single glazed. Space heating is supplied by means of a gas boiler and radiators in the rooms, while DHW is supplied by single-apartment electric boilers in 5 apartments and by a gas boiler in one apartment. Lighting is obtained by incandescent lamps. With reference to the TRNSYS typical meteorological year, the annual use of primary energy for heating, DHW and lighting is 332.5 kWh/m^2 and the annual emission of CO_2 is 92.5 kg/m^2. The efficiency targets of the retrofitting, prescribed in the HERB DoW, are a reduction of at least 80 % in the use of primary energy, a reduction of at least 60 % in the CO_2 emission, an annual use of primary energy less than 50 kWh/m^2, excluding appliances. The proposed retrofitting includes: external thermal insulation of the vertical walls by a 16 cm layer of calcium silicate hydrates; insulation of loft and floors towards unheated or external spaces by mineral wool; replacement of windows; installation of a multifunction air-to-water heat pump for heating, cooling and DHW; replacement of the radiators by high-efficient fan coils and low temperature radiators; LED lighting; installation of PV panels on the Southeast and the Northeast pitches of the roof. The following efficiency targets are expected to be reached: 86.5 % reduction of the use of primary energy, 86.3 % reduction of CO_2emission, annual use of primary energy equal to 44.8 kWh/m^2, including summer cooling and dehumidifying, presently unavailable.

2. DESCRIPTION OF THE BUILDING AND CLIMATIC DATA

The demonstration house is located in an area very close to the center of Bologna, main city of the region Emilia-Romagna, North-Center Italy. It is composed of three floors, with 2 apartments each, an unheated attic, and an unheated basement which contains the boiler room. The roof has 4 pitches, with orientations Southeast, Southwest, Northeast and Northwest. The first floor is larger than the second and the third, which are identical. The house has a small garden, with high trees. Street views of the Northeast and of the Southwest side of the house are reported in Fig. 4; 3-D models of the house, which illustrate the Northeast and Northwest sides and the Southwest and Southeast sides, are reported in Fig. 5. In the 3-D models, most trees have been hidden.

Apartments have been numbered as follows: Apartments 1a and 1b at the first floor; Apartments 2a and 2b at the second floor; Apartments 3a and 3b at the third floor. Apartments with letter a are facing Northwest; those with letter b are facing Southeast. The first floor is larger than the second and the third. One of the apartments at the first floor (Apartment 1b) has, as a part of the roof, an inaccessible plane terrace. The second and the third floor are identical. Plans of the apartments are reported in Fig. 6: the first floor is represented on the left, the second and the third floor (which are identical) are represented on the right. The rectangle which appears in the lower part of the figure at right represents the inaccessible terrace which forms a plane roof for a part of Apartment 1b. The heated floor areas of the apartments are as follows: Apartment 1a: 44.7 m^2; Apartment 1b: 80.2 m^2; Apartments 2a and 3a: 39.2 m^2; Apartments 2b and 3b: 39.3 m^2. The total heated floor area is 281.9 m^2.

Figure. 4. Street views of the house: Northeast side (left) and Southwest side (right)

Figure. 5. 3-D models of the house: Northeast and Northwest sides (left); Southwest and Southeast sides (right)

The external wall, made of solid brick masonry, is 31 cm thick and uninsulated. Most windows are single glazed, with wood frame, except those of Apartments 2b and 3b, which are double glazed with aluminum frame. The internal height of each floor is 3.36 m, and the floor thickness is 0.27 m. The unheated basement is placed under Apartment 1a and under about one third of Apartment 1b. The rest of Apartment 1b is on the ground. No thermal insulation is placed between the first floor and the basement, or the ground, as well as between Apartment 1b and the terrace. The unheated loft is non ventilated and uninsulated; its lowest height is 0.20 m (at the sides), and its highest height is 1.60 m (at the center). It has a very small window, which is closed.

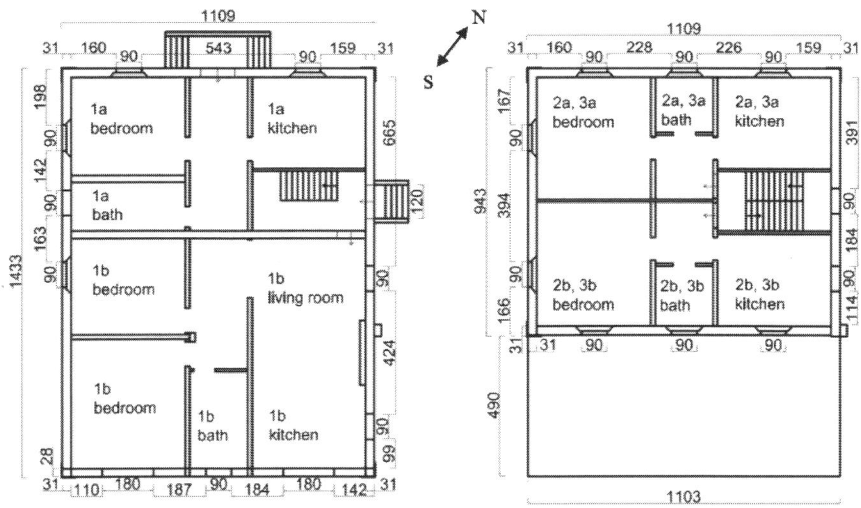

Figure. 6. Plans of the apartments: first floor on the left; second and third floor on the right

Space heating is supplied by means of a gas boiler, placed in the basement, and radiators in the rooms, under windows. The gas boiler is a modulating one, installed in 2007; it has a nominal power of 62 kW, a certified efficiency equal to 0.93 at nominal power and slightly higher at reduced power. A constant efficiency equal to 0.93 has been considered. The distribution, emission and control efficiencies of the heating system can be assumed as equal to 0.90, 0.95 and 0.97 respectively, according to the national standard UNI TS 11300–2. Thus, the overall efficiency of the heating system is 0.771.

DHW is supplied by single apartment boilers: electric boilers in Apartments 1a, 2a, 2b, 3a, 3b; a gas boiler in Apartment 1b. No summer air conditioning is present. All rooms, including bathrooms, have windows.

The main properties of the opaque enclosure elements in the present state (decription, thickness *s*, transmittance *U*), are summarized in Table 1; for each element, the description starts from the internal (heated) space. Heat transfer coefficients have been determined by considering standard thermal conductivities of the materials employed, without final roundations: there is no claim of accuracy in the last decimal places.

Table 1. Main properties of the opaque enclosure elements, in the present state

Element	Description	s [m]	U [W/ (m^2K)]
Vertical wall	Internal plaster, solid brick masonry, external plaster	0.31	1.797
Floor on basement	Tiles, subfloor, floor brick and concrete, plaster	0.26	1.476
Floor on ground	Tiles, subfloor, floor brick and concrete	0.25	1.845
Terrace floor	Plaster, floor brick and concrete, subfloor, tiles	0.26	1.476
Loft floor	Internal plaster, floor brick and concrete	0.09	2.903
Roof	Internal plaster, floor brick and concrete	0.21	2.279

Bologna is located in the center of the region Emilia Romagna, at the border between the Padana flat and the foot of the mountains Appennino Tosco-Emiliano. The mean altitude is 54 m above sea level, the latitude is 44° 29' North, the longitude is 11° 20' East. Since the dynamic simulations of the building are performed through TRNSYS 17, the typical meteorological year (TMY) for Bologna available in that program is employed. The mean monthly values of the external air temperature T_e obtained from the TRNSYS TMY, listed in the first row of Table 2, are slightly lower than those reported in the National Standard UNI 10349, which are listed in the second row of Table 2. The values of the monthly mean relative humidity according to TRNSYS TMY are reported in the third row of Table 2, and those according to UNI 10349 are reported in the fourth row. The average wind speed, according to UNI 10349, is 1.6 m/s, with main direction Southwest. The mean monthly values of the beam radiation H_b and of the diffuse radiation H_d on a horiontal surface according to UNI 10349, in MJ/m^2 per day, are illustrated in Fig. 7.

3. SIMULATION SCENARIOS

The following simulation scenarios have been considered.

1 Present State

2 Retrofit 1 (thermal insulation of walls and floors)

3 Retrofit 2 (thermal insulation of walls and floors, replacement of windows)

4 Retrofit 3 (thermal insulation of walls and floors, replacement of windows, installation of an air-to-water heat pump for heating and DHW)

5 Retrofit 4 (thermal insulation of walls and floors, replacement of windows, installation of an air-to-water heat pump for heating and DHW, LED lighting)

6 Retrofit 5 (thermal insulation of walls and floors, replacement of windows, installation of an air-to-water heat pump for heating, cooling, dehumidifying and DHW, LED lighting, PV collectors).

Table 2. Monthly mean values of temperature and relative humidity in Bologna,

TRNSYS (first and third row) and UNI 10349 (second and fourth row)

Month	Jan	Feb	March	April	May	June	July	Aug	Sept	Oct	Nov	Dec
T_e [°C]	1.7	4.3	9.4	13.8	20.2	21.5	24.4	24.1	20.9	14.4	8.4	3.9
T_e [°C]	2.1	4.6	9.4	14.2	18.2	22.9	25.4	24.9	21.2	14.9	8.7	4
φ	0.82	0.78	0.71	0.73	0.71	0.70	0.66	0.68	0.71	0.75	0.83	0.82
φ	0.89	0.87	0.72	0.67	0.68	0.65	0.57	0.60	0.64	0.74	0.85	0.86

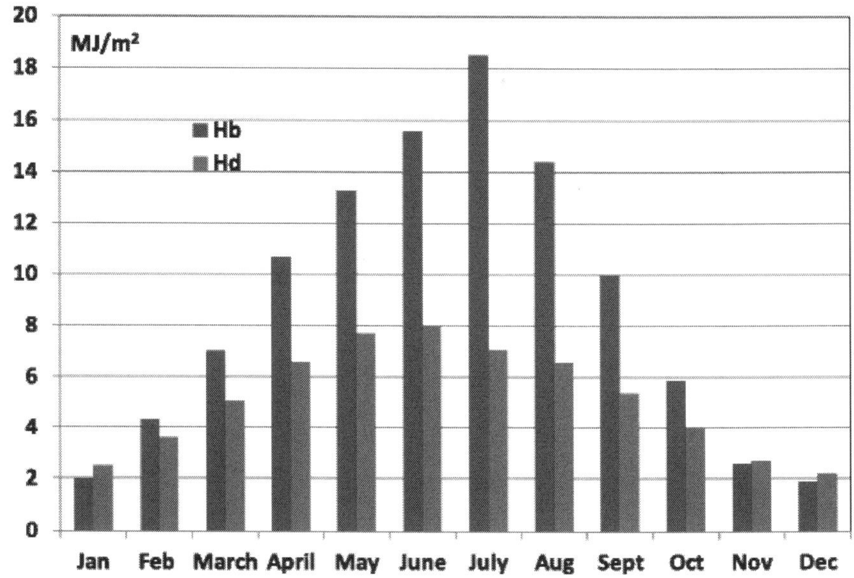

Figure. 7. Mean monthly values of direct and diffuse radiation on a horizontal surface, per day

In Scenario Retrofit 1, the following insulations of the opaque enclosure components are adopted. The vertical wall is insulated externally by a 16 cm layer of calcium silicate hydrates, with thermal conductivity 0.045 W/(mK). This material is recyclable and requires a very low use of primary energy for its production. The floor on basement is insulated by a 9.5 cm layer of mineral wool, with thermal conductivity 0.032 W/(mK). The terrace between first and second floor is insulated by a 14 cm layer of mineral wool, with thermal conductivity 0.037 W/(mK). The loft floor is insulated by a 20 cm layer of mineral wool, with thermal conductivity 0.04 W/(mK). The floor on ground and the roof are unchanged. The main properties of the retrofitted enclosure elements are reported in Table 3; for each element, the description starts from the internal (heated) space.

Besides the thermal insulation of the opaque enclosure elements, scenario Retrofit 2 includes the replacement of all windows by double glazed windows with wood frame, having a 16 mm Argon layer and low emissivity glass; the glass transmittance is $U_g = 1.00$ W/(m^{2}K) and frame transmittance is $U_f = 1.84$ W/(m^{2}K).

Table 3. Main properties of the retrofitted enclosure elements

Element	Description	s [m]	U [W/(m^{2}K)]
Vertical wall	Internal plaster, solid brick masonry, external plaster, hydrate calcium silicates 16 cm	0.47	0.243
Floor on basement	Tiles, subfloor, floor brick and concrete, plaster, mineral wool 9.5 cm	0.355	0.274
Terrace floor	Plaster, floor brick and concrete, subfloor, tiles, mineral wool 14 cm	0.40	0.224
Loft floor	Internal plaster, floor brick and concrete, mineral wool 20 cm	0.29	0.187

In addition to the retrofitting elements of Retrofit 2, Scenario Retrofit 3 includes the installation of a multifunction air-to-water heat pump with inverter, which provides heating and DHW. The capacity of the thermal storage for DHW is 1.0 m^3; that of the thermal storage for heating is 0.2 m^3. In order to ensure a high COP of the heat pump during the heating season and to allow also summer cooling and dehumidifying (additional service provided only in Scenario Retrofit 5), the present radiators are replaced by new high-efficiency fan coils for living rooms and bedrooms, and by low temperature radiators for bathrooms. The new fan coils and radiators operate, during the heating season, with a water inlet temperature between 40 °C and 38 °C. Values of the thermal power supplied by the heat pump in heating mode, with water delivered at 40 °C and return temperature 34 °C, for several values of the external air temperature T_e and of the compressor frequency, are reported in Table 4; the corresponding values of the COP are reported in Table 5.

When the heat pump works for DHW production, water is delivered with highest temperature 50 °C and lowest temperature 48 °C. The corresponding return temperatures are 40 °C and 38 °C. Values of the COP of the heat pump working for DHW production with evaporator in external air and water delivered at 50 °C, for several values of the external air temperature and of the compressor frequency, are reported in Table 6.

Table 4. Heating power of the heat pump, kW, with hot water delivered at 40 °C

	Frequency [Hz]				
T_e [°C]	110	90	70	50	30
−7	7.63	6.01	4.53	3.15	1.89
2	9.99	7.91	6.01	4.19	2.51
7	11.70	9.14	6.95	4.90	2.93
12	13.50	10.60	8.09	5.66	3.41

Table 5. COP of the heat pump, with hot water delivered at 40 °C

	Frequency [Hz]				
T_e [°C]	110	90	70	50	30
−7	2.65	2.71	2.67	2.51	2.15
2	3.27	3.38	3.39	3.21	2.78
7	3.75	3.85	3.88	3.72	3.22
12	4.29	4.44	4.49	4.30	3.76

Table 6. COP of heat pump for DHW production at 50 °C and evaporator in external air

	Frequency [Hz]				
T_e °C	110	90	70	50	30
−7	2.19	2.22	2.19	2.06	1.79
2	2.64	2.71	2.70	2.57	2.25
7	2.97	3.06	3.04	2.94	2.57
12	3.36	3.49	3.49	3.37	2.97
20	4.15	4.35	4.36	4.24	3.74
25	4.69	4.89	4.93	4.76	4.22
30	5.21	5.44	5.47	5.29	4.66
35	5.71	5.97	6.00	5.81	5.14

In addition to the retrofitting elements of Retrofit 3, Scenario Retrofit 4 includes the replacement of incandescent light bulbs by LED lighting. Besides the retrofitting elements of Retrofit 4, Scenario Retrofit 5 includes the installation of PV collectors with a conversion efficiency 14.5 % in the Southeast pitch and in the Northeast pitch of the roof: an area of 21.45 m^2 is installed in the Southeast pitch; an area of 7.8 m^2 is installed in the Northeast pitch. The peak power of the PV system is 4.24 kW. The installation of PV panels in the Southwest pitch is not convenient because this pitch is shadowed by high trees. In Scenario Retrofit 5, the heat pump provides also summer cooling and dehumidifying. When the heat pump works in cooling/dehumidifying mode and heat for DHW is required, the condensation heat is supplied to the DHW thermal storage. In the cooling/dehumidifying mode, cold water is delivered at 7 °C and returns at 12 °C. Values of the EER of the heat pump with condenser in external air, for several values of the external air temperature T_e and of the compressor frequency, are reported in Table 7. Values of the EER of the heat pump with condensation heat supplied to DHW, either at T_{hw} = 50 °C or at T_{hw} = 48 °C, are reported in Table 8.

Table 7. EER of heat pump with cold water delivered at 7 °C and condenser in external air

	Frequency [Hz]				
T_e [°C]	110	90	70	50	30
20	4.78	5.22	5.43	5.34	4.62
25	4.08	4.45	4.63	4.56	3.98
30	3.48	3.8	3.96	3.92	3.45
35	2.97	3.24	3.39	3.36	3.00

4. BUILDING ENERGY SIGNATURE IN THE PRESENT STATE

In order to validate the TRNSYS simulation model and the technical data collected for the building in the Present State, a pre-monitoring of the building started in May 2013 and is ongoing. The validation of the simulation code has been performed by considering the measured values of the internal air temperature, of the external air temperature and of the gas volume used for heating, with reference to the period from October 15th 2013 to April 15th 2014. Hourly values of the internal air temperature were measured by sensors placed in two rooms per apartment (kitchen and bedroom). Hourly values of the external air temperature were taken from measurements performed by the Urbana Weather Station, Bologna, close to the building. The volume of the gas used for heating was obtained through weekly readings of the gas meter connected with the central boiler. The values of the internal air temperature for Apartment 3a are reported in Fig. 8, as an example. The mean temperature for this apartment, in the period considered, was 20.3 °C for the berdroom and 20.9 °C for the kitchen. The mean internal-air temperature for the whole building, averaged among all apartments, was 20.2 °C, which can be considered as equal to the set temperature for the internal air, namely 20 °C. The values of the external air temperature, during the same period, are reported in Fig. 9 and compared with those of TRNSYS TMY. The figure shows that winter 2013–2014 was much milder than usual: the mean temperature from October 15th 2013 to April 15th 2014 was 10.15 °C, while that of TRNSYS TMY is 6.99 °C. A similar anomaly is happening in winter 2014–2105.

Table 8. EER of heat pump with cold water delivered at 7 °C and heat supplied to DHW

	Frequency [Hz]				
T_{hw} [°C]	110	90	70	50	30
48	2.45	2.59	2.63	2.54	2.23
50	2.30	2.42	2.47	2.40	2.11

The data employed to determine the experimental energy signature of the building are reported in Table 9. The duration of each period, in hours, is not always a multiple of 24 h, due to differences in the reading times, ranging from 1:30 pm to 5:00 pm. The lower heating value of the natural gas supplied to the building is 9.7 kWh/m^3. The equation of the energy signature obtained is

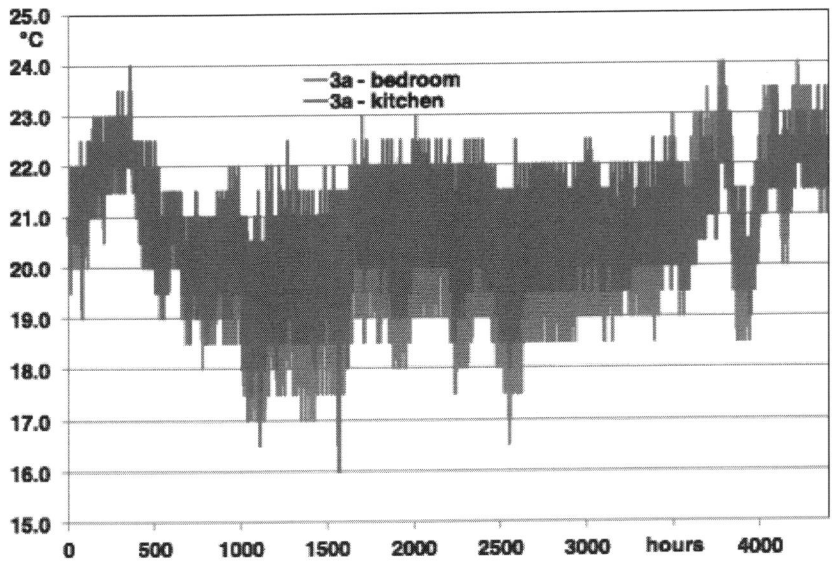

Figure. 8. Internal air temperature in Apartment 3a, from October 15th 2013 to April 15th 2014

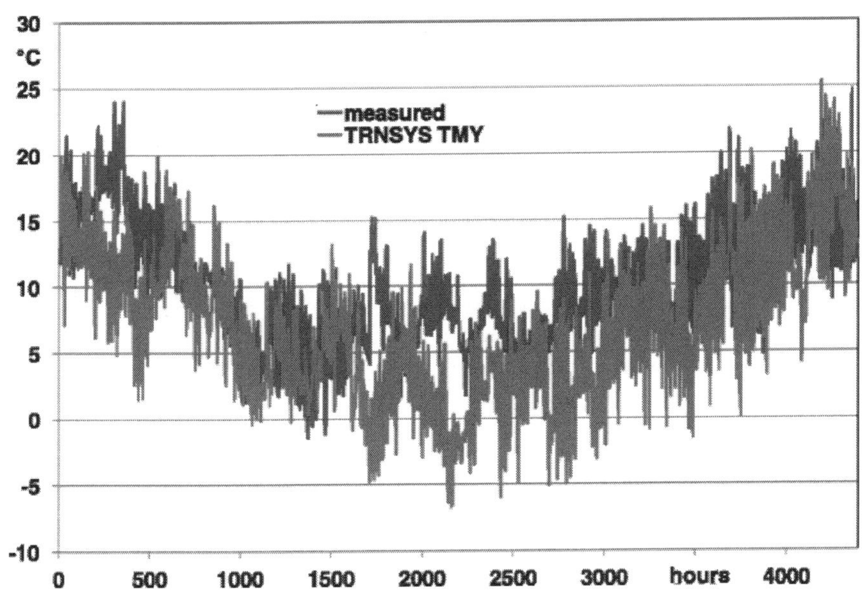

Figure. 9. Measured external air temperature compared with TRNSYS TMY, from October 15th 2013 to April 15th 2014

Table 9. Data employed to determine the experimental energy signature

Period	Hours	Volume m^3	Energy kWh	Power kW	T_e °C
Oct. 15 – Oct. 30	360	110	1067	2.96	17.23
Oct. 30 – Nov. 30	744	866	8400.2	11.29	10.68
Nov. 30 – Dec. 12	288	584	5664.8	19.67	5.1
Dec. 12 – Jan. 20	936	1585	15374.5	16.43	7
Jan. 20 – Jan. 27	168	273	2648.1	15.76	7.91
Jan. 27 – Feb. 03	168	327	3171.9	18.88	5.18
Feb. 03 – Feb. 10	167.5	263	2551.1	15.23	8.64
Feb. 10 – Feb. 17	167.5	242	2347.4	14.01	9.13
Feb. 17 – Feb. 24	167	224	2172.8	13.01	9.24
Feb. 24 – March 03	169	236	2289.2	13.55	9.34
March 03 – March 10	170.5	214	2075.8	12.17	10.43
March 10 – March 17	167	165	1600.5	9.58	12.34
March 17 – March 24	168.5	135	1309.5	7.77	13.66
March 24 – March 31	166.5	171	1658.7	9.96	11.88
March 31 – April 07	170	108	1047.6	6.16	15.35
April 07 – April 15	190.5	74	717.8	3.77	16.1

$$\dot{Q}_{\text{exp}} = 26.209 - 1.3541\, T_e, \tag{1}$$

where Q˙expQ˙exp is the mean value of the primary energy use per unit time, in kW, and T_e is the mean value of the external air temperature in °C.

The data employed to determine the energy signature of the building in the Present State by the dynamic simulation performed through TRNSYS 17 are reported in Table 10. The period considered is from October 1st to April 30th of the TRNSYS TMY, with intervals of half month. The primary energy use in each period has been determined by dividing the thermal energy need, calculated through the dynamic simulation, by the overall efficiency of the heating plant. The latter has been set equal to 0.795, namely the product of the boiler efficiency 0.93, the distribution efficiency 0.9, and the emission efficiency 0.95. In the comparison between the experimental energy signature and the computational one, the control efficiency has been set equal to 1, because the real average value of the internal air temperature was 20 °C. The equation of the computational energy signature for the Present State is

$$\dot{Q}_{\text{comp}} = 27.575 - 1.6208\, T_e, \tag{2}$$

where Q˙compQ˙comp is the mean value of the primary energy use per unit time, in kW, obtained throgh the dynamic simulation.

A comparison of the experimental energy signature and the computational one is illustrated in Fig. 10. The comparison shows that the gross heat loss coefficient of the building (namely the heat loss coefficient which takes into account also the plant efficiency) determined through the dynamic simulation, 1.62 kW/K, is 19 % higher that that found experimentally. Raesons for this discrepancy could be: a lower air exchange rate in the real operation with respect to that considered in the dynamic simulation (0.3 volumes per hour); a difference between the real heat loss coefficient and that calculated, for some enclosure elements; a higher plant efficiency with respect to that considered; an inaccuracy of the gas volume meter. Another difference between the experimental energy signature and the computational one is the higher value of the zero-load temperature which occurs in the real case. This difference is probably due to the lower internal heat gains in the real case with respect to the standard ones.

Table 10. Data employed to determine the computational energy signature

Period	Hours	Energy kWh	Power kW	T_e °C
Oct. 1 – Oct. 15	360	404.7	1.12	16.62
Oct. 16 – Oct. 31	384	2668.2	6.95	12.39
Nov. 01 – Nov. 15	360	4169.6	11.58	10.30
Nov. 16 – Nov. 30	360	6109.5	16.97	6.48
Dec. 01 – Dec. 15	360	7732.7	21.48	4.57
Dec. 16 – Dec. 31	384	8513.4	22.17	3.24
Jan. 01 – Jan. 15	360	8997.3	24.99	1.15
Jan.16 – Jan. 31	384	9409.0	24.50	2.23
Feb. 01 – Feb. 15	360	8374.7	23.26	2.37
Feb. 16 – Feb. 28	312	5568.1	17.85	6.61
March 01 – March 15	360	5582.1	15.51	6.93
March 16 – March 31	384	3313.0	8.63	11.74
April 01 – April 15	360	1431.7	3.98	15.04
April 16 – April 30	360	1835.0	5.10	12.61

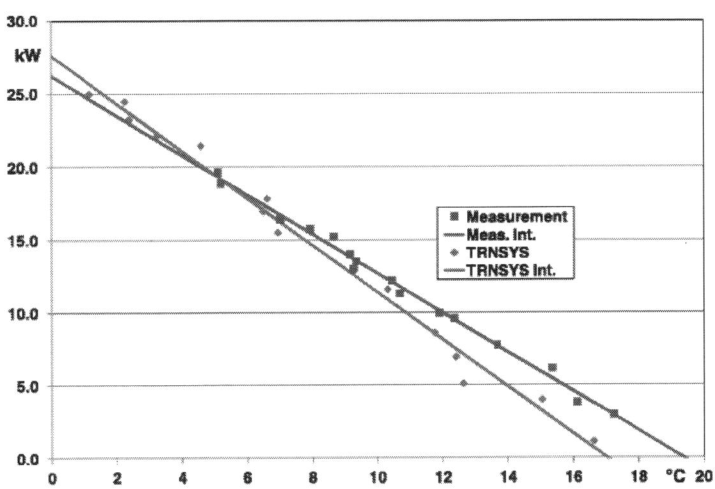

Figure. 10. Comparison between the experimental energy signature and the computational one

In fact, during the period considered most of the tenants were often outside and had a small use of energy for cooking and lighting. A more detailed analisis of the comparison between the experimental energy signature and the computed one will be performed in a specific paper. Here we will assume that the comparison illustrated in Fig. 10 is an acceptable validation of the simulation code and of the technical data collected on the building in its present state.

In the following, the annual use of primary energy in each scenario will be considered as equal to that obtained through the dynamic simulation.

5. ANNUAL USE OF PRIMARY ENERGY IN EACH RETROFIT SCENARIO

The dynamic simulation of the building in each retrofit scenario was performed through TRNSYS 17. The dynamic simulation of the air-to-water heat pump, with gas-boiler auxiliary, was performed by two own MATLAB codes: one for operation in heating and DHW production mode (with heating from October 1st to April 30th); one for operation in cooling/dehumidifying and/or DHW production mode (with cooling and dehumidifying from May 1st to September 30th). The evaluation of the electric energy produced by the PV system was performed according to the National Standards UNI/TS 11300–4 (2012) and UNI/TR 11328–1 (2009)

6. PRIMARY ENERGY USE FOR HEATING AND DHW IN THE PRESENT STATE AND SCENARIOS RETROFIT 1 AND RETROFIT 2

The dynamic simulations through TRNSYS 17 allowed to determine the annual energy need for heating of the building, in each retrofit scenario, for each thermal zone and, as a consequence, for each apartment and for the whole building. The values of the annual energy need for heating in kWh, in the Present State and in Scenarios Retrofit 1 and Retrofit 2 are reported in Table 11, for each apartment and for the whole building. The energy need for heating in Scenarios Retrofit 3, Retrofit 4 and Retrofit 5 coincides with that in Scenario Retrofit 2. A hourly plot of the thermal power required by the building for heating, in Scenarios Retrofit 2 ÷ Retrofit 5, is reported in

Fig. 11; it shows that the thermal power required by the building, in a TRNSYS 17 typical meteorological year, never exceeds 10 kW.

Table 11. Annual energy need for heating in each retrofit scenario, kWh

Scenario	Ap. 1a	Ap. 1b	Ap. 2a	Ap. 2b	Ap. 3a	Ap. 3b	Building
Present State	9487	17976	7479	6245	9475	8391	59,054
Retrofit 1	4579	9111	3435	2375	3867	2868	26,234
Retrofit 2 ÷ 5	2873	6114	1773	1608	2190	2112	16,670

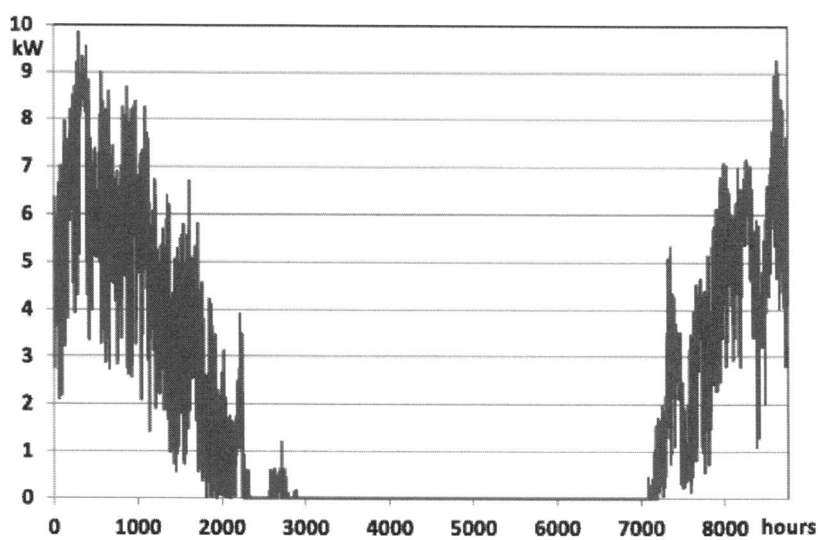

Figure. 11. Thermal power required by the building for heating, in Scenarios Retrofit 2 ÷ Retrofit 5

The primary energy use for heating in the Present State and in Scenarios Retrofit 1 and Retrofit 2 is determined by dividing the corresponding thermal energy need by the product of the following coefficients: boiler efficiency (0.93), distribution efficiency (0.90), emission efficiency of the radiators (0.95), control efficiency (0.97). The product of these coefficients is 0.771. The values of the primary energy use for heating in these scenarios are reported in Table 12.

Table 12. Annual use of primary energy for heating in Present State, Retrofit 1 and Retrofit 2, kWh

Scenario	Ap. 1a	Ap. 1b	Ap. 2a	Ap. 2b	Ap. 3a	Ap. 3b	Building
Present State	12,301	23,306	9697	8096	12,285	10,879	76,564
Retrofit 1	5937	11813	4453	3079	5014	3718	34,013
Retrofit 2	3725	7927	2299	2085	2839	2738	21,614

The thermal energy need for DHW is determined by applying the National Standard UNI/TS 11300–2, where the hot water is assumed to be delivered at 40 °C. The annual need for the single apartments and for the building, in kWh, is reported in Table 13, first row of results.

Table 13. Annual energy need for DHW, and use in Present State, Retrofit 1, Retrofit 2, kWh

	Ap. 1a	Ap. 1b	Ap. 2a	Ap. 2b	Ap. 3a	Ap. 3b	Building
Need	853.3	1367.4	748.3	750.2	748.3	750.2	5217.8
Use	2771	2188	2430	2436	2430	2436	14,689

In the Present State and in Scenarios Retrofit 1 and Retrofit 2, DHW is supplied by an electric boiler in each apartment, except Apartment 1b, where a gas boiler is employed. According to UNI TS 11300–2, the emission and distribution efficiencies can be assumed as equal to 0.95 and 0.94, respectively. For an electric boiler, the efficiency (including thermal losses) can be assumed equal to 0.75. Since the efficiency of the electricity production system in Italy is 0.46, the overall efficiency for the electric boilers of Apartments 1a, 2a, 2b, 3a, 3b (including emission and distribution) is 0.308. The efficiency of a DHW gas boiler (including thermal losses) can be assumed equal to 0.70; therefore, the overall efficiency (including emission and distribution) for the gas boiler of Apartment 1b is 0.625. The values of the primary energy use for DHW in Present State and in Scenarios Retrofit 1 and Retrofit 2 are reported in Table 13, second row of results.

The values of the primary energy use for heating and DHW, in the Present State and in Scenarios Retrofit 1 and Retrofit 2 are reported in Table 14.

Table 14. Annual use of primary energy for heating and DHW in Scenarios Present State, Retrofit 1 and Retrofit 2, kWh

Scenario	Ap. 1a	Ap. 1b	Ap. 2a	Ap. 2b	Ap. 3a	Ap. 3b	Building
Present State	15,071	25,494	12,127	10,532	14,715	13,315	91,254
Retrofit 1	8707	14,001	6883	5515	7444	6154	48,703
Retrofit 2	6495	10,115	4729	4521	5269	5174	36,303

7. PRIMARY ENERGY USE FOR HEATING AND DHW IN SCENARIOS RETROFIT 3 AND RETROFIT 4

In Scenarios Retrofit 3 and Retrofit 4, heating and DHW are provided by a multifunction air to water heat pump and the present radiators are replaced by high efficient fan coils (one for each room) and low-temperature radiators (in bathrooms). New distribution circuits for heating and DHW are installed. The distribution, emission and control efficiencies for heating can be assumed as equal to 0.98, 0.98, 0.98. The emission and distribution efficiencies for DHW can be assumed equal to 0.95 and 0.96, respectively. Relevant technical data of the heat pump are reported in Tables 4, 5 and 6. The amounts of electric energy used by the heat pump for heating and DHW from October 1st to April 30th have been determined by a MATLAB hourly simulation code. The monthly amounts of the electric energy used by the heat pump during that period, in kWh, are reported in the first row of Table 15; those used by the fan coils and the total amounts are reported in the second and in the third row. The amounts of electric energy used by the heat pump for DHW from May 1st to September 30th are reported in Table 16.

Table 15. Electric energy used by heat pump and fan coils from October 1st to April 30th, in Scenarios Retrofit 3 and Retrofit 4, kWh

	Oct	Nov	Dec	Jan	Feb	March	April	Season
Heat pump	258	869	1485	1704	1253	692	275	6536
Fan coils	8	15	26	26	19	16	8	117
Total	266	884	1511	1730	1272	708	283	6653

Table 16. Electric energy used by heat pump for DHW from May 1st to September 30th, in Scenarios Retrofit 3 and Retrofit 4, kWh

	May	June	July	Aug	Sept	Season
Electric	133	123	118	120	125	619

The total annual use of electric energy for heating and DHW is 7272 kWh. By considering the efficiency of the electricity production system in Italy, 0.46, one finds a corresponding use of primary energy of 15809 kWh. The primary energy used by the gas boiler, for integration during October÷May, is 414 kWh. Therefore, the total annual use of primary energy for heating and DHW in Scenarios Retrofit 3 and Retrofit 4 is 16223 kWh. By partitioning this use between the apartments, in proportion to their needs, one obtains the results reported in Table 17.

8. PRIMARY ENERGY USE FOR LIGHTING

An experimental study performed by Politecnico di Milano in 2004 (Di Andrea and Danese 2004) showed that the average annual use of electric energy for lighting in Italy, for apartments with an average area of 106 m^2, is 375 kWh, i.e., 3.54 kWh/m^2. The apartments considered in Ref. (Di Andrea and Danese 2004) had 80 % incandescent lamps, 15.8 % fluorescent lamps, and 4.2 % halogen lamps, a composition similar to that of the apartments under exam here, in scenarios Present State ÷ Retrofit 3. We assume that for apartments 1a, 2a, 2b, 3a, 3b, with an area of about 50 m^2, the energy use per unit area is 20 % higher, i.e., 4.25 kWh/m^2. With this assumption, the annual use of electric energy for lighting in kWh, for each apartment, is reported in Table 18. The first row refers to Scenarios Present State ÷ Retrofit 3, while the second row refers to Scenarios Retrofit 4 and Retrofit 5, where all lamps are replaced by LED lighting. The corresponding use of primary energy for lighting is reported in Table 19, where the efficiency 0.46 of the Italian electricity system has been considered.

Table 17. Annual use of primary energy for heating and DHW in Scenarios Retrofit 3 and Retrofit 4, kWh

Ap. 1a	Ap. 1b	Ap. 2a	Ap. 2b	Ap. 3a	Ap. 3b	Building
2762	5545	1869	1748	2178	2121	16,223

Table 18. Annual use of electric energy for lighting, kWh

Scenario	Ap. 1a	Ap. 1b	Ap. 2a	Ap. 2b	Ap. 3a	Ap. 3b	Building
Present State ÷ Retrofit 3	190	284	167	167	167	167	1141
Retrofit 4 and Retrofit 5	38	57	33	33	33	33	228

9. PRIMARY ENERGY USE FOR HEATING, DHW AND LIGHTING, IN SCENARIOS PRESENT STATE ÷ RETROFIT 4

The total annual use of primary energy for heating, DHW and lighting, in Scenarios Present State ÷ Retrofit 4, in kWh, is resumed in Table 20. The primary energy use per unit area, in kWh/m^2, is reported in the last column. The table shows that Scenario Retrofit 4 yields a percent reduction of primary energy use equal to 82.2 % with respect to the Present State.

Table 19. Annual use of primary energy for lighting, kWh

Scenario	Ap. 1a	Ap. 1b	Ap. 2a	Ap. 2b	Ap. 3a	Ap. 3b	Building
Present State ÷ Retrofit 3	413	617	362	363	362	363	2481
Retrofit 4 and Retrofit 5	83	123	72	73	72	73	496

Table 20. Annual use of primary energy for heating, DHW and lighting in Scenarios Present State ÷ Retrofit 4, kWh and kWh/m^2

Scenario	Ap. 1a	Ap. 1b	Ap. 2a	Ap. 2b	Ap. 3a	Ap. 3b	Building	per m^2
Present State	15,484	26,111	12,489	10,895	15,077	13,678	93,734	332.5
Retrofit 1	9120	14,618	7245	5878	7806	6517	51,184	181.6
Retrofit 2	6908	10,733	5091	4884	5631	5537	38,784	137.6
Retrofit 3	3175	6162	2231	2111	2540	2484	18,703	66.3
Retrofit 4	2844	5669	1941	1820	2250	2194	16,719	59.3

10. PRIMARY ENERGY USE FOR HEATING, COOLING, DEHUMIDIFYING, DHW AND LIGHTING, IN SCENARIO RETROFIT 5

To further reduce the annual use of primary energy and to provide also summer cooling and dehumidifying, Scenario Retrofit 5 is considered, where PV panels are installed on the roof of the building: an area of 21.45 m^2 is installed in the Southeast pitch; an area of 7.8 m^2 is installed in the Northeast pitch. The amounts of electric energy used by heat pump and fan coils from October 1st to April 30th are the same as in Scenarios Retrofit 3 and Retrofit 4, and are reported in Table 15. The primary energy used by the gas boiler, in the same period, is 414 kWh.

Cooling and dehumidifying is provided from May 1st to September 30th: the set point is 27 °C for the internal air temperature and 50 % for the relative humidity. The monthly amounts of thermal energy need for cooling and dehumidifying for the whole building, evaluated by dynamic simulation through TRNSYS 17, are reported in Table 21. Also the monthly amounts of the thermal energy use are reported in the last row, obtained by considering an overall distribution, emission and control efficiency equal to 0.98 × 0.98 × 0.98.

Table 21. Monthly amounts of the thermal energy need and use for cooling and dehumidifying, kWh

	May	June	July	Aug	Sept	Season
Sensible	566	1309	2350	2110	855	7190
Latent	102	175	286	273	125	962
Total	669	1485	2636	2383	980	8153
Total use	710	1577	2801	2532	1041	8662

The monthly amounts of the electric energy used by the heat pump for cooling, dehumidifying and DHW production during the period from May 1st to September 30th were determined by applying a hourly MATLAB simulation code. The results, in kWh, are reported in Table 22. In the same table, also the estimated use of electric energy by the fan coils is reported. The primary energy used by the gas boiler for integration of DHW during this period is about 1 kWh.

Table 22. Monthly amounts of the electric energy used for cooling, dehumidifying and DHW from May 1st to September 30th, kWh

	May	June	July	Aug	Sept
Heat pump	353	526	824	779	423
Fan coils	8	15	26	26	8
Total	361	541	850	805	430

The monthly values of the total electric energy used (for heating, cooling-dehumidifying, DHW and lighting) are reported in the first row of Table 23. Part of the electric energy used is provided by the PV system. Monthly amounts of the electric energy produced by the PV system are reported in the second row of the table. In the other rows, the following monthly energy amounts are reported, in kWh: electric energy taken from the grid, corresponding primary energy, PV electric energy employed for self use, PV electric energy supplied to the grid. The table shows that the annual use of primary energy due to the use of electric energy from the grid is 12215 kWh. By adding the primary energy used by the gas boiler, 415 kWh, one obtains the total annual use of primary energy of the building, 12630 kWh, which corresponds to 44.8 kWh/m^2.

11. SYNOPSIS OF THE TOTAL ANNUAL USE OF PRIMARY ENERGY IN SCENARIOS PRESENT STATE ÷ RETROFIT 5

A synopsis of the total annual use of primary energy per unit heated floor area in Scenarios Present State ÷ Retrofit 5 is given in Fig. 12. The proposed retrofit scenario is Retrofit 5, which yields 86.5 % saving of primary energy with respect to the Present State and provides also summer cooling and dehumidifying, a service not available in other scenarios.

Table 23. Monthly electric energy balance and primary energy use in Scenario Retrofit 5, kWh

	Jan	Feb	Mar	April	May	June	July	Aug	Sept	Oct	Nov	Dec	Year
TOTAL	1749	1289	727	301	381	560	870	825	449	285	903	1530	9869
PV	148	224	356	467	564	603	681	577	431	306	167	139	4663
From grid	1601	1066	371	0	0	0	189	247	18	0	736	1391	5619
Primary	3481	2317	806	0	0	0	410	537	39	0	1600	3024	12,215
Self use	148	224	356	301	381	560	681	577	431	285	167	139	4251
To grid	0	0	0	166	183	43	0	0	0	21	0	0	413

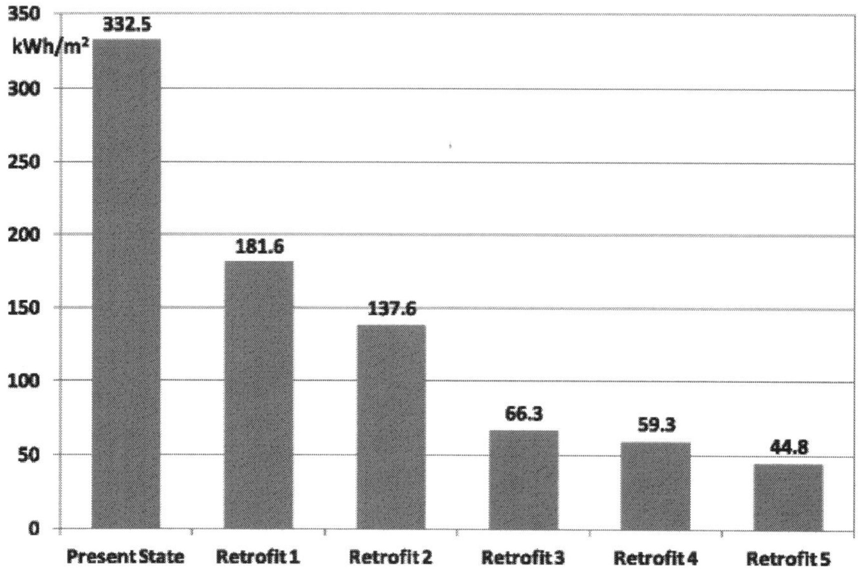

Figure. 12. Total annual use of primary energy per unit floor area in all Scenarios

12. ANNUAL EMISSION OF CO_2 IN EACH RETROFIT SCENARIO

The emission of carbon dioxide caused by the use of fossil fuels and of electric energy can be calculated by employing the European Standard EN 15603:2008. The fuel employed in the building is natural gas. The CO_2 production coefficient is 277 kg/MWh for natural gas and 617 kg/MWh for electric energy.

The annual use of primary thermal energy for heating (THE heat), of electric energy for heating (EE heat), of primary thermal energy for DHW (THE DHW), of electric energy for DHW (EE DHW), of electric energy for lighting (EE light), of total primary thermal energy (THE), of total electric energy (EE), in kWh are reported in columns 2 ÷ 8 of Table 24. In columns 9 ÷ 11 of the same table are reported the kilograms of CO_2 due to the use of thermal energy (CO_2 THE), the kilograms of CO_2 due to the use of electric energy (CO_2 EE), the total kilograms of CO_2 emitted per year.

Table 24. Annual use of primary thermal energy and of electric energy, kWh, and annual emission of CO_2, kg

Scenario	THE heat	EE heat	THE DHW	EE DHW	EE light	THE	EE	CO_2 THE	CO_2 EE	CO_2
Present State	76,564	0	2187	5749	1141	78,751	6890	21,814	4251	26,065
Retrofit 1	34,013	0	2187	5749	1141	36,201	6890	10,028	4251	14,279
Retrofit 2	21,614	0	2187	5749	1141	23,801	6890	6593	4251	10,844
Retrofit 3	414	6653	0	619	1141	414	8413	115	5191	5305
Retrofit 4	414	6653	0	619	228	414	7500	115	4628	4742
Retrofit 5	414		1			415	5619	115	3467	3582

In Scenarios Retrofit 3 and Retrofit 4, the use of electric energy for heating (EE heat) includes that for the production of DHW during the heating period. In Scenario Retrofit 5, the total use of electric energy cannot be divided into use for heating (EE heat) and use for DHW (EE DHW), on account of the PV electricity input. The table shows that the reduction in CO_2 emission obtained in Scenario Retrofit 5 is 86.3 % with respect to the Present State.

13. COMFORT IMPROVEMENT

A study on the thermal comfort has been carried out, to estimate the comfort improvement produced by the retrofitting. Dynamic simulations performed through TRNSYS 17 allowed to determine the time evolution of operative temperature, of the relative humidity, and of the Fanger indexes PMV (Predicted Mean Vote) and the PPD (Predicted Percentage Dissatisfied) (Fanger 1970) for each room, in scenarios Present State and Retrofot 5, with reference to TRNSYS TMY. The PMV index predicts the mean response of a large group of people according to a standard thermal sensation scale, ranging from +3 (hot) to −3 (cold), where 0 stands for neutral. The PPD index is related to the PMV index. Usual recommended limits for PMV and PPD indexes are $-0.5 < PMV < 0.5$ and $PPD < 20$ %, respectively. With reference to the standard UNI EN ISO 7730:2006, the following input conditions have been considered in winter and summer, respectively. For the heating period: clothing insulation of 1.0 clo (typical business suit), metabolic rate of 1.2 met (seated, light home work), air speed of 0.1 m/s; for the cooling period: clothing insulation of 0.5 clo (light summer clothing), metabolic rate of 1.2 met (seated, light home work), air speed of 0.3 m/s.

The results show that a relevant improvement of the thermal comfort can be obtained by the retrofitting, especially during summer, and that in scenario Retrofit 5 the PPD is expected to remain below the recommended limit of 20 % for every room, during the whole year. Some results are illustrated in Figs. 13, 14, 15 and 16, with reference to Apartment 3b, kitchen.

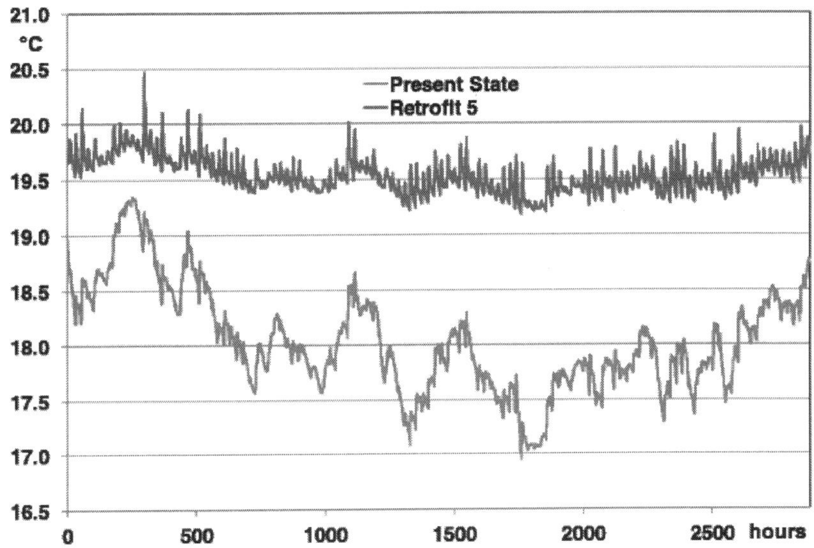

Figure. 13 Operative temperature for Apartment 3b, kitchen, from November 1st to February 28th

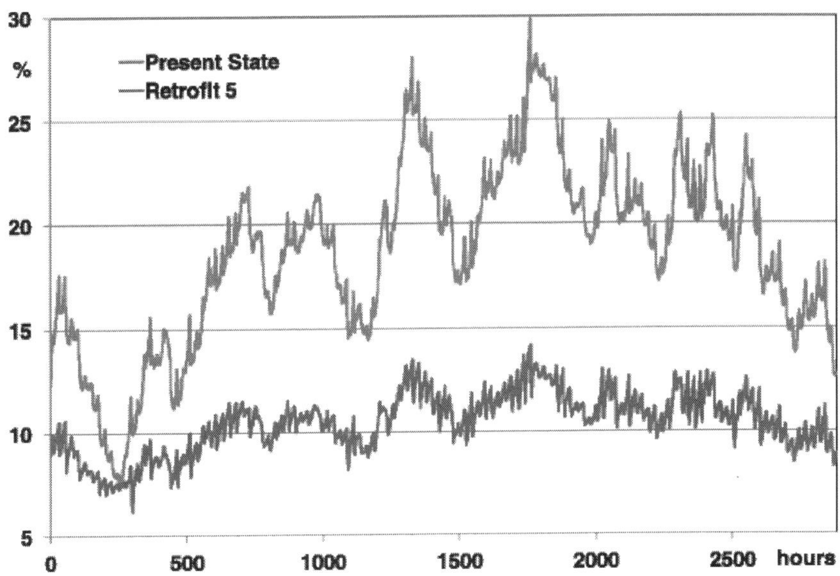

Figure. 14. PPD for Apartment 3b, kitchen, from November 1st to February 28th

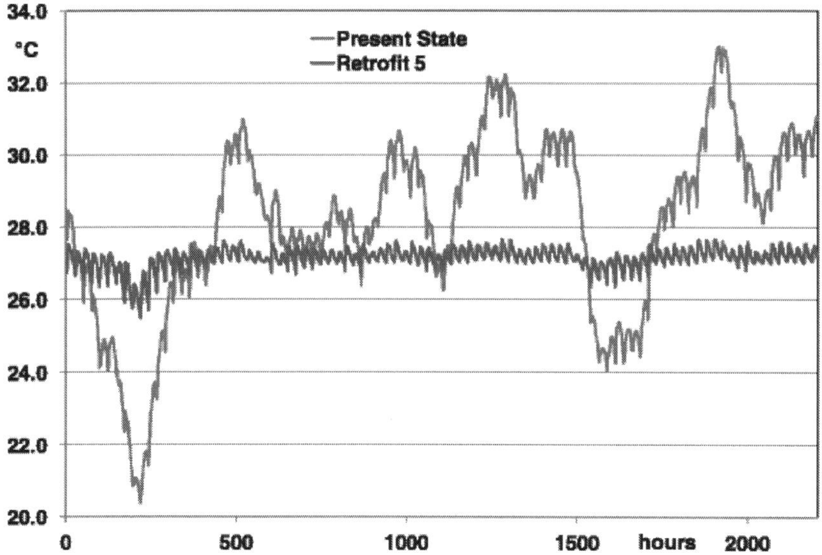

Figure. 15. Operative temperature for Apartment 3b, kitchen, from June 1st to August 31st

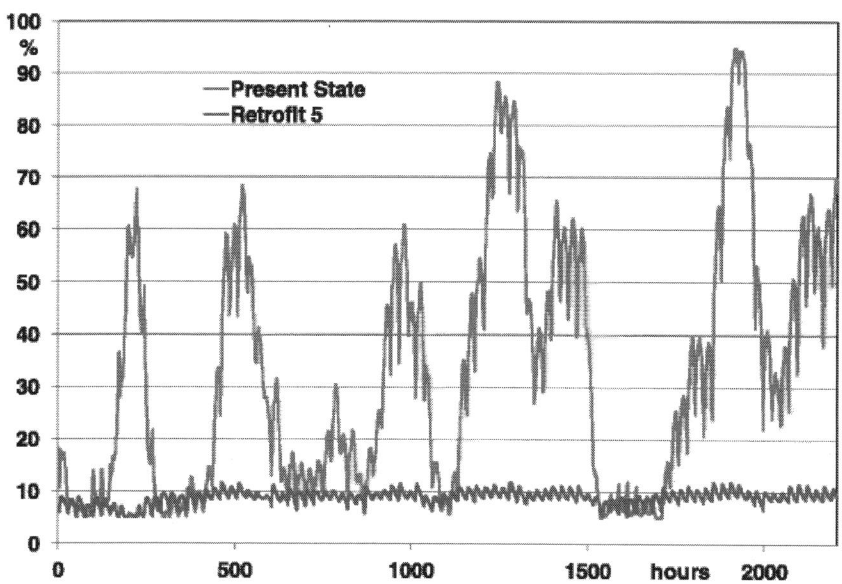

Figure. 16. PPD for Apartment 3b, kitchen, from June 1st to August 31st

Figures 13 and 14 illustrate, respectively, the time evolution of the operative temperature and of the PPD during winter, from November 1st to February 28th. Fig. 13 shows that, while in the Present State the operative emperature is often lower than 18 °C, in Retrofit 5 it is very close to the set temperature of the internal air, 20 °C. Fig. 14 shows the corresponding reduction of PPD. Figs. 15 and 16 illustrate, respectively, the time evolution of the operative temperature and of the PPD during summer, from June 1st to August 31st. Fig. 15 evidences that the combined effects of the thermal resistance and of the thermal inertia of the thermal insulation keep the operative temperature of the room very stable, even if the external temperature suddenly decreases and remains low for some days. Fig. 16 shows that the PPD, dramatically high in the Present State, remains below 10 % in Retrofit 5.

14. ECONOMIC ANALYSIS

For each retrofit scenario, the additional cost with respect to the previuos scenario and the total cost have been determined. Then, for each scenario, the annual money saving due to the reduction of use of energy sources has been determined. Hence, for each retrofit scenario, the financial payback time at zero interest rate, zero annual increase of the cost of energy sources, zero inflation rate has been calculated.

The additional cost and the total cost of each retrofit scenario are summarized in Table 25. The additional cost of Retrofit 3 (replacement of the plant for heating and DHW) is very high because, for the building considered, the replacement of the plant requires a complete construction (or reconstruction) of the distribution circuit. In fact, no distribution circuit for DHW is present, and the distribution circuit for heating is uninsulated and obsolete. Indeed, the very high additional cost of Retrofit 3 could be considered only in part as a cost for energy retrofitting, and in part as a necessary renovation cost of the building. The additional cost for the installation of heat pump, thermal storage tanks, fan coils and radiators is 24,555 Euro; the remaining additional cost (114,045 Euro) is for the installation of the new distribution circuits for heating and DHW and of heat meters.

Table 25. Additional cost and total cost of each retrofit scenario, Euro

Retrofit scenario	Add cost	Total cost
Retrofit 1	67,419	67,419
Retrofit 2	18,324	85,743
Retrofit 3	138,600	224,343
Retrofit 4	871	225,214
Retrofit 5	11,660	236,874

In order to determine the annual economic savings, the following data on the employed fuel, natural gas, and on electricity have been employed:

- Natural gas: lower heating value 9.595 kWh/m^3; cost 0.9 Euro/m^3;
- Electricity: cost 0.2 Euro/kWh; value of electricity supplied to grid 0.0389 Euro/kWh

The value of the PV electricity supplied to grid is very low at present in Italy. This is a sudden change with respect to the policy of a few years ago, when very high incentives were given to PV energy production and also the electricity produced and self-used was paid.

The annual savings obtainable in each retrofit scenario are illustrated in Table 26. The columns of the table report, for each scenario: the annual use of thermal primary energy (kWh), the annual use of electric energy (kWh), the annual use of natural gas (m^3), the cost of the gas used (Euro), the cost of the electricity used (Euro), the total annual cost of gas plus electricity (Euro), the income due to PV electricity supplied to grid (Euro), the annual saving with respect to the Present State (Euro).

In the evaluation of the payback time, for each retrofit scenario, State incentives should be considered. Although the building is owned by the Municipality of Bologna, reference is made to the Italian State incentives delivered to private citizens, as recommended by the HERB consortium. The retrofit interventions are considered in the following sequence: thermal insulation of vertical and horizontal walls (Retrofit 1); Retrofit 1 + replacement of windows (Retrofit 2); Retrofit 2 + replacement of the heating and DHW plant (Retrofit 3); Retrofit 3 + LED lighting (Retrofit 4); Retrofit 4 + PV system (Retrofit 5). In this sequence, each additive cost receives a State contribution equal to 65 % of the cost. The LED lighting and the PV system receive this contribution because they are parts of a general retrofit plan which yields an annual use of primary energy lower than the limit prescribed by the Government for the applicability of the 65 % contribution. The State contribution is given in the form of tax reduction in 10 equal amounts, during a period of 10 years. Since vanishing interest and inflation rates are assumed, the total State contribution is considered as a cost reduction available at the first year.

Table 26. Use of energy sources and savings in each scenario, per year

Scenario	THE kWh	EE kWh	Gas m^3	Gas cost	EE cost	Cost	Income	Saving
Present State	78,751	6890	8119	7307	1378	8685	0	—
Retrofit 1	36,201	6890	3732	3359	1378	4737	0	3948
Retrofit 2	23,801	6890	2454	2208	1378	3586	0	5098
Retrofit 3	414	8413	43	38	1683	1721	0	6964
Retrofit 4	414	7500	43	38	1500	1538	0	7146
Retrofit 5	415	5619	43	39	1124	1162	16	7523

The total cost and the total payback time of each retrofit scenario, without incentives and with incentives, are reported in Table 27. As shown in the table, the replacement of the plant, with the installation of new distribution circuits, enhances the general payback time with incentives from about 6 to about 11 years. This intervention, however, yields several benefits to the building: the possibility of supplying heating, DHW, cooling and dehumidifying with a very low annual use of primary energy; the renovation of the distribution system.

15. CONCLUSIONS

Five scenarios for the energy retrofitting of a residential building with 6 apartments located in Bologna, North-Center Italy, have been analyzed through hourly dynamic simulations of the building and of the plant. Simulations of the building have been performed by TRNSYS 17; those of the multifunction heat pump system with thermal storage by own MATLAB codes. The simulation code has been validated, in the Present State scenario, by comparing the computational energy signature of the building with that determined experimentally by a pre-retrofit monitoring.

Table 27. Total cost (Euro) and total payback time (years) of each retrofit scenario

Scenario	Cost	Cost with incentives	Payback	Payback with incentives
Retrofit 1	67,419	23,597	17.1	6.0
Retrofit 2	85,743	30,010	16.8	5.9
Retrofit 3	224,343	78,520	32.2	11.3
Retrofit 4	225,214	78,825	31.5	11.0
Retrofit 5	236,874	82,906	31.5	11.0

The proposed retrofit scenario is the most complete one, which includes: thermal insulation of the external vertical walls by calcium silicate hydrates; thermal insulation of horizontal enclosure elements by mineral wool; replacement of windows; installation of a multifuction heat pump with inverter and thermal storage for heating, cooling/dehumidifying, and DHW supply; installation of new distribution networks for heating/cooling and DHW; installation of high efficient fan coils and low temperature radiators; LED lighting; installation of a PV system on the roof. The proposed retrofit scenario yields a 86.5 % reduction of the use of primary energy, a 86.3 % reduction of CO_2 emission, and a relevant comfort improvement. The payback time for private owners, considering State incentives, is 11 years.

ACKNOWLEDGMENT

This research was funded by the European Union's Seventh Framework Program, Theme EeB.NMP.2012-2, Project HERB (Holistic energy-efficient retrofitting of residential buildings), Grant agreement no: 314283.

AUTHORS' CONTRIBUTIONS

EZ coordinated the research work, evaluated the energy use and the CO_2 production in each scenario, by employing simulation results provided by SL and CN, and wrote the paper. CN performed the dynamic simulations of the heat pump system. SL performed the dynamic simulations of the building. GLM evaluated the plant efficiency in the present state and gave suggestions for the choice of the retrofit scenarios. All authors read and approved the final manuscript.

REFERENCES

1. Afonso C, Oliveira AC (2000) Solar chimneys: simulation and experiment. Energy Build 32:71–79
2. Amstalden RW, Kost M, Nathani C, Imboden DM (2007) Economic potential of energy-efficient retrofitting in the Swiss residential building sector: The effects of policy instruments and energy price expectations. Energy Policy 35:1819–1829
3. Ascione F, de Rossi F, Vanoli GP (2011) Energy retrofit of historical buildings: theoretical and experimental investigations for the modelling of reliable performance scenarios. Energy Build 43:1925–1936
4. Balaras CA, Droutsa K, Argiriou AA, Asimakopoulos DN (2000) Potential for energy conservation in apartment buildings. Energy Build 31:143–154
5. Bizzarri G, Morini GL (2006) New technologies for an effective energy retrofit of hospitals. Appl Therm Eng 26:161–169
6. Brignoli R, Cecchinato L, Zilio C (2013) Experimental analysis of an air–water heat pump with micro-channel heat exchanger. Appl Therm Eng 50:1119–1130
7. Chan HY, Riffat SB, Zhu J (2010) Review of passive solar heating and cooling technologies. Renew Sust Energ Rev 14:781–789

8. Chidiac SE, Catania EJC, Morofsky E, Foo S (2011a) A screening methodology for implementing cost effective energy retrofit measures in Canadian office buildings. Energy Build 43:614–620

9. Chidiac SE, Catania EJC, Morofsky E, Foo S (2011b) Effectiveness of single and multiple energy retrofit measures on the energy consumption of office buildings. Energy 36:5037–5052

10. Cuce E, Cuce PM, Wood CJ, Riffat SB (2014) Optimizing insulation thickness and analysing environmental impacts of aerogel-based thermal superinsulation in buildings. Energy Build 77:28–39

11. Dall'O' G, Sarto L (2013) Potential and limits to improve energy efficiency in space heating in existing school buildings in northern Italy. Energy Build 67:298–308

12. Daouas N (2011) A study on optimum insulation thickness in walls and energy savings in Tunisian buildings based on analytical calculation of cooling and heating transmission loads. Appl Energy 88:156–164

13. Di Andrea F, Danese A (2004) Misure dei consumi di energia elettrica nel settore domestico., http://www.eerg.it/resource/pages/it/Progetti_-_MICENE/compendio_misure_consumi_elettrici.pdf

14. Energy-efficient buildings. Multiannual roadmap for the contractual PPP under Horizon 2020. http://www.ectp.org/cws/params/ectp/download_files/36D2981v1_Eeb_cPPP_Roadmap_under.pdf

15. Eurostat, European Commission portal for statistics. http://ec.europa.eu/eurostat/data/database

16. Fanger PO (1970) Thermal comfort: Analysis and applications in environmental engineering. Copenhagen (Denmark): Danish Technical Press

17. International Energy Statistics – EIA (US Energy Information Administration). http://www.eia.gov/cfapps/ipdbproject/IEDIndex3.cfm#

18. IPCC Fifth Assessment Report – AR5. https://www.ipcc.ch/report/ar5/wg1/

19. Liu X, Ni L, Lau SK, Li H (2013) Performance analysis of a multi-functional heat pump system in heating mode. Appl Therm Eng 51:698–710

20. Naldi C, Morini GL, Zanchini E (2014) A method for the choice of the optimal balance-point temperature of air-to-water heat pumps for heating. Sustain Cities Soc 12:85–91

21. Orosa JA, Oliveira AC (2009) Energy saving with passive climate control methods in Spanish office buildings. Energy Build 41:823–828

22. Pisello AL, Santamouris M, Cotana F (2013) Active cool roof effect: impact of cool roofs on cooling system efficiency. Adv Build Energy Res 7:209–221

23. Santamouris M (2014) Cooling the cities – a review of reflective and green roof mitigation technologies to fight heat island and improve comfort in urban environments. Sol Energy 103:682–703

24. Santamouris M, Synnefa A, Karlessi T (2011) Using advanced cool materials in the urban built environment to mitigate heat islands and improve thermal comfort conditions. Sol Energy 85:3085–3102

25. Shahrokni H, Levihnc F, Brandt N (2014) Big meter data analysis of the energy efficiency potential in Stockholm's building stock. Energy Build 78:153–164

26. Terlizzese T, Zanchini E (2011) Economic and exergy analysis of alternative plants for a zero carbon building complex. Energy Build 43:787–795

27. Ucar A, Balo F (2009) Effect of fuel type on the optimum thickness of selected insulation materials for the four different climatic regions of Turkey. Appl Energy 86:730–736

28. Yu J, Yang C, Tian L, Liao D (2009) A study on optimum insulation thicknesses of external walls in hot summer and cold winter zone of China. Appl Energy 86:2520–2529

29. Zhao X, Lee JM, Riffat SB (2008) Numerical study of a novel counter-flow heat and mass exchanger for dew point evaporative cooling. Appl Therm Eng 28:1942–1951

30. Zhao X, Yang S, Duan Z, Riffat SB (2009) Feasibility study of a novel dew point air conditioning system for China building application. Build Environ 44:1990–1999

Chapter 4

A simple and effective model for prediction of effective thermal conductivity of vacuum insulation panels

Ankang Kan, Liyun Kang, Chong Wang and Dan Cao

Merchant Marine College, Shanghai Maritime University

ABSTRACT

The excellent thermal insulation performance of vacuum insulation panels (VIPs) make them widely applied in energy conservation fields, especially in buildings engineering. This research work proposes a simple, yet extremely effective, alternative model for prediction of the effective thermal conductivity (ETC.) of VIPs. The ETC. of VIPs is function of the thermal conductivity of the core materials, the equivalent thermal conductivity of the rarefied gas embraced in the core and the equivalent thermal conductivity of radiation in the early studies. The micro structure of the porous core materials and vacuum degree are taken into consideration and the prediction numerical model for the ECT of VIPs is developed. The relationship of the vacuum degree versus the ETC. is theoretical analyzed. Three types VIPs are made from the polyurethane foam materials, superfine fibrous materials and nano-granular silica materials as the core materials. For each type, the vacuum degree and the thermal conductivity are collected, including the comparison between the testing results and the prediction model. The agreement between the model and the experimental results is fairly well when the air pressure is very low. The vacuum maintaining and service life of the VIPs are also discussed. The research work is meaningful for the enhancement of stability and the development of vacuum insulation panels.

KEYWORDS

Vacuum insulation panels, Thermal conductivity, Vacuum degree, Numerical analysis and prediction

1. INTRODUCTION

Nowadays, vacuum insulation panels (VIPs) are regarded as one of the optimum thermal adiabatic materials for the energy conservation purpose on the market. The thermal insulation performances observed, even ten times better than common heat insulation materials can be achieved by applying VIPs, resulting in a great potential for the reduction of energy loss in thermal space with slim exterior-protected walls (Fricke et al. 2008; Baetens 2010; Nussbaumer et al. 2006; Nussbaumer et al. 2005; Brunner et al. 2012; Caps et al. 2001). The flat VIPs contain a porous core material which withstands the atmospheric air pressure, a gas-tight barrier envelope that is optimized for low air & moisture leakage rate and for a long service life to maintain the internal vacuum level, and getter or gas absorption materials, if necessary, to absorb internal gas from leakage or other sources. That is, the VIPs make use of vacuum to suppress heat transfer due to rarefied gaseous conduction (Bouquerel et al.2012a; Bouquerel et al. 2012b; Alam et al. 2011). So the thermal adiabatic property of VIPs dramatically depends on the core materials and the internal gas pressure. The porous materials, such as, micro-open foam, nano-structured power and fine fiber, are commonly used as core materials of VIPs, which are easily evacuated and have minimum thermal conduction effect (Kwon et al. 2009). The heat transfer processing in VIPs is via solid conduction, rarefied gas conduction, radiation and convection occurring at interface of gas and solid wall which is nearly zero, is always ignored in calculation. The solid conduction and radiation depends on the structure, the porosity and properties of core materials, while the gas conduction by the residual gases embraced in the caves depends on the internal gas pressure which maybe increased with time. So the ETC. of VIPs must be function of the variables mentioned above.

Various models to predict the ETC. of VIPs can be founded in the literature. The typical models for VIPs with different core materials have been presented by Jae Sung Kwon *et al.* (Kim & Song 2013; Caps & Fricke 2000; Di et al. 2013; Di et al. 2014). Special emphasis on the solid conduction is theoretically investigated and transport mechanisms of three core materials are numerically analyzed. Separate calculation for solid conductivity, gaseous conductivity and radiative conductivity is also studied. However, separate research of these contributions is not easy in practice because any of them cannot be fully eliminated. There are also many variables in the models that should be determined by experiments. Based on an empirical approach, Kan and Han (2013) proposed a fractal model to estimate the ETC. of open-cell polyurethane foam. This model consists of fractal dimension and fractal diameter which are not easily obtained. Wang (Wang & Pan 2008a; Wang & Pan 2008b; Wang et al. 2007) presents a lattice Boltzmann method, that is, a random generation-growth method to produce micro morphology of porous media. Different parameters of different porous materials are involved in the statistical models and the simulation porous materials are self-similarity in the micro space. A simplified model are proposed by Miguel A. A. Mendes *et al.* (Mendes et al. 2013; Mendes et al.2014) to estimate the ETC. of open-cell foam-like porous materials and the serial and parallel arrangements are simulated in the model. They concluded that the ETC. strongly depends on the porosity and the ratio of thermal conductivities of solid and fluid phases. The results are compared with the experimental data, and the correlation for the model is also obtained.

From the considered models mentioned above, it can be easily concluded that the numerical prediction models, where the effective structural cell is generated based on the real morphology of the porous media, are the function of the porosity, the thermal conductivities of the phases, the radiation conductivity and additional empirical parameters in some case. However, such models can be found in the references, but they do not directly provide explicit expressions for the ETC. of VIPs because of the vacuum conditions. As the wide application of VIPs, the needs of the ETC. prediction are urgent for manufacture, development and research. In view of above discussion, the main objective of this study is to provide an effective model to predict the ETC. of VIPs with alterable core materials. Based on the morphology of the core materials and the vacuum degree, the appropriate simplified numerical models are investigated for alterable VIPs.

2. MATHEMATICAL METHODS

For the general porous insulation materials under the normal condition, the heat transfer flux is composed of four parts, according to the traditional heat transfer theory, that is, heat conduction by the solid matrix itself λ_s, heat conduction by the filled phases (mainly gas, vapor or other fluent) in the cavities λ_g, heat convection between solid phase and filled phases λ_c and radiation equivalent thermal conduction among the phases λ_y. Neglecting the coupling effect by all the phases, the effective thermal conductivity of the porous media can be express as:

$$\lambda_e = f(\lambda_s, \lambda_g, \lambda_c, \lambda_r) \tag{1}$$

However, for VIPs, the air pressure in the cavities of the core materials is very small, nearly zero. So the fluent phase is always gas, under the normal temperature. The thermal convection between the solid wall and the filled phases can be ignored.

2.1 The conductivity of solid matrix

The heat conduction for the nonmetal solid matrix can be achieved by the vibration of the lattice morphology. So the certain porous materials, the ETC. is usually regarded as the function of the mean temperature. In fact, the ETC. also has the relationship with the materials density, the micro structure, the characteristics, the porosity, and so on. The calculation formula is so very complex that it is nearly impossible to get the accuracy value. So the value is always provided by vendors or collected from experiment.

2.2 The conductivity of rarefied gas

Free motion of gas molecules makes the heat transfer complex in the open cells of the core porous materials. According to the gas kinetic theory, we can acknowledge that, the smaller the cavity dimension, the harder the gas molecules move in it. For the VIPs core materials, the pore dimension scales are micron or even nanometer level. While the pore dimension is no larger than 100 μm and the air pressure is under 100 Pa, that is, the Reynolds is smaller than 1000 and Knudsen number is bigger than ten, the equivalent conductivity of gas can be ignored, $\lambda_c \approx 0$.

Based on the rarefied gas heat transfer theory, the ETC. of the rarefied gas embraced in the cavities is given by (Kwon et al.2009; Kan et al. 2013):

$$\lambda_g = \frac{\lambda_0}{1 + 2\beta K_n} \tag{2}$$

Where: λ_0 is the ETC. of static gas under the normal temperature and normal air pressure condition, W/(m · K), here $\lambda_0 = 0.0230$ W/(m · K); β is a constant number, which is used to express the grade the gas molecules collide the solid walls, and its value is always between 1.5 and 2, depending on the gas type, core material characteristics and mean temperature, here $\beta = 1.5$; K_n, Knudsen number, is the ratio of the gas molecule mean free path l_m and the mean diameter of the pore δ, that is $K_n = l_m/\delta$; and l_m is determined by (Kwon et al. 2009):

$$l_m = \frac{K_B T}{\sqrt{2}\pi d_g^2 p_g} \tag{3}$$

Where T is the mean temperature of the porous material in thermodynamic scale, K; d_g is the diameter of the gas molecule, m, here, for air, $d_g = 3.72 \times 10^{-10}$ m; K_B is the Boltzmann constant, and $K_B = 1.38 \times 10^{-23}$ m J/K; P_g is the rarefied air pressure, Pa.

Combined Eqs. (2) and (3), the ECT of rarefied gas embraced in pore can be obtained as the follow:

$$\lambda_g = \frac{\lambda_0}{1 + \frac{\sqrt{2}\beta K_B T}{\pi d_g^2 P_g \delta}} \tag{4}$$

2.3 The equivalent thermal conductivity of radiation

Thermal radiation, in the form of electromagnetic wave for energy transfer, can occur without any medium, even in the vacuum case. And the equivalent thermal conductivity of radiation can be calculated by the Eq. (5) (Mendes et al. 2013):

$$\lambda_r = 4l_c\sigma(T_1 + T_2)(T_1^2 + T_2^2)/3\phi \tag{5}$$

Where, l_c is the thickness of the plate core materials, m; φ is the attenuation coefficient for porous media, and here $\varphi = 445\ m^{-1}$;$\sigma$is Steve Boltzmann constant, and $\sigma = 5.6697 \times 10^{-8}\ W/(K^4 \cdot m^2)$; T_1,T_2 are respectively both sides of the VIPs temperature in thermodynamic scale, K.

2.4 The ETC. of VIPs

To calculate the ETC. of the VIPs, the explicit mathematical model is derived from Eq. (1) to yield the following Eq. (6):

$$\lambda_e = (1-\xi)\lambda_s + \xi\lambda_g + \lambda_r \tag{6}$$

Where, ξ is the porosity of the core materials.

Combined with Eqs. (4), (5) and (6), the ETC. of the VIPs can be expressed as the following:

$$\lambda_e = (1-\xi)\lambda_s + \xi\frac{\lambda_0}{1+\frac{\sqrt{2}\beta K_B T}{\pi d_g^2 P_g \delta}} + \frac{4l_c\sigma(T_1 + T_2)(T_1^2 + T_2^2)}{3\phi} \tag{7}$$

3. NUMERICAL ANALYSIS AND DISCUSSION

3.1 Porous materials and simulation

Three typical porous media, that is, open cell foam structure, granular porous structure and fibrous porous structure, as the core material for VIPs, are involved in the simulation. The micro structures are shown in Fig. 1.

a b c

Figure. 1. The micro profiles of three typical porous materials as VIPs core materials. (**a**) micro section of open cell foam structure (**b**) micro section of fibrous porous structure (**c**) micro section of granular silicon dioxide porous structure

The thickness of VIPs, assuming is 20 mm, is employed in the simulation. And the simulation boundary conditions of the ETC. testing, which involves the heat guarded method, is that the hot plate temperature is 35 °C while the cold side is 15 °C, taking the GB/T3399-2009 as the simulating reference. The core porous materials, the open cell polyurethane which has the thermal conductivity of the solid phase 0.5832 W/(m · K), the pore size range is 1-100 μm and the porosity is 95 %, superfine glass fiber with the thermal conductivity of the solid phase 1.3 W/(m · K),the diamter range 100 nm-100 μm and the porosity 95 %, the nano-grain silica with the thermal conductivity of the solid phase is 0.27 W/(m · K),the diameter range 10 nm-10 μm and porosity 95 %, are taken in the theoretical calculation. The results are illustrated in Fig. 2.

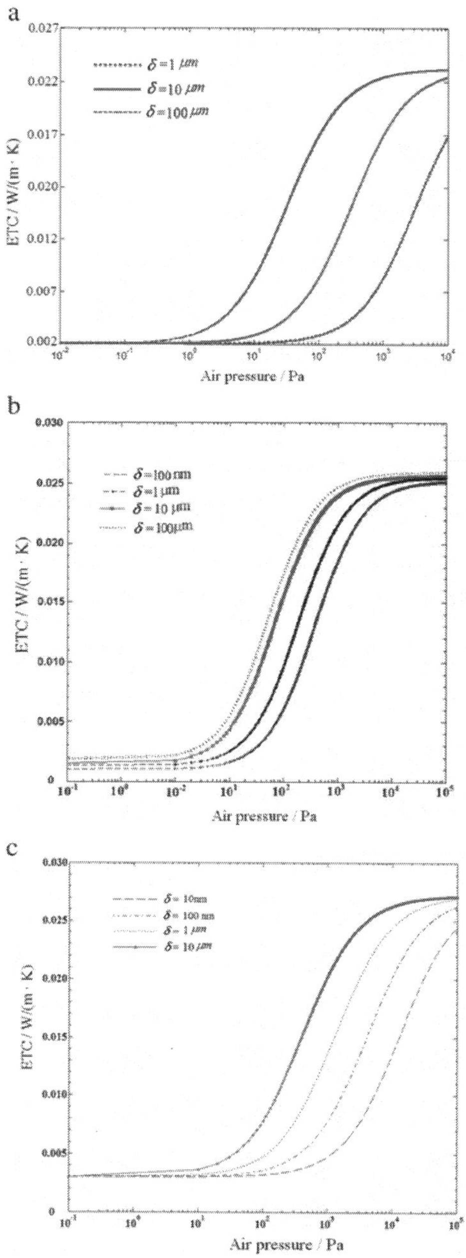

Figure. 2. a The vacuum degree in open micro cell polyurethane and the pore diameters versus the ETC. in simulation **b**) The vacuum degree in superfine glass fiber and the fiber diameters versus the ETC. in simulation **c**) The vacuum degree in nano-grain silica dioxide and the pore diameters versus the ETC. in simulation

4. RESULTS AND DISCUSSION

Barrier membrane with good air-blocking and heat-sealing properties, gas getters and driers putting inside the core media, are helpful for the vacuum maintenance. Combined with the physical properties parameters of porous media, the vacuum degree inside the VIPs and the gas permeability ratio of barriers, the service life of the VIPs can be modeled and predicted.

- The relationship of the vacuum degree versus the ETC.

Figure 2 shows the predicted ETC. of three typical porous media under vacuum condition. The curves in the simulation for the open micro cell polyurethane, superfine glass fiber and nano-grain silica dioxide are very similar. In the relevant simulation figure, the curve is nearly flat while the static air pressure in the cavities is over a certain value P_{low}, that is, the ETC. is nearly constant. But there is a sharp decline of the ETC. with the reducing of the static air pressure while the vacuum value is less than the certain value P_{low} mentioned above. When the static air pressure in the cavities reaches for another certain value P_{low}, the air in the cells is so rarefied that thermal convection can be negligible, and the curve line is flat again with the air pressure decreasing. For the VIP manufacture and application, the certain air pressure P_{low} is meaningful and critical. When the air pressure inside the porous core materials is below the critical P_{low}, the VIP will take a good adiabatic thermal performance in the usage. Taking into account the information in Fig. 2, one can easily make the conclusion that, if one wants to get the ideal insulating VIPs, the air pressure inside the core materials, for corresponding micro cellular polyurethane, should be reduced to below 1 Pa, for superfine glass fiber, below 10 Pa, for the nano-grain silica dioxide, below 100 Pa.

The critical factor for the thermal properties and thermal stability of VIP is the vacuum degree inside the core materials, as can be seen from Fig. 2. So the vacuum maintenance, that is, below the P_{low}, is essential and vital for the manufacturing and application of the VIPs. Water, embraced in cavities of the core materials, will change into vapor when the air pressure is very low at the normal temperature. And the vapor, if presents, will deteriorate the thermal property of VIPs. Therefore, core materials pre-treating proceeding for the manufacturing VIPs is important for the vacuum maintaining through the service life. The barrier membrane and the numerical model in this paper, the service life of VIPs can be deduced for further research work (Kim et al.2012; Schwab et al. 2005a; Schwab et al. 2005b; Jung et al. 2013; Tenpierik & Cauberg 2007).

- The relationship of the feature size of the porous medium versus the ETC.

From the Fig. 2, one can also easily recognize that, the ETC. greatly reduces with the reduction of the feather size of the porous core materials, as the occupancy of solid matrix increases and the corresponding air molecules free path decreases, and the movement of air embraced in the pore is restricted in the narrow spaces. With the feather size decreasing, the porosity decreases but the corresponding P_{low} increased. In practice, the improvement of the P_{low} is meaningful for the special equipments, especially for the vacuum heat sailing machines. But it is also the obstacles for the vacuum degree to create.

Actually, the recommending vacuum value for the open cellular polyurethane, as the core materials, is within 100 μm, for super fine glass fiber, within 10 μm, and super nano- grain silica dioxides within 100 nm.

5. VIPS SAMPLES AND EXPERIMENTS

5.1 Samples of VIPs with different core materials

Three typical porous core materials were selected and the same making craft was taken to make the VIPs samples, in order to compare the texting ETC. with that of the simulation results. And every five samples were made with different inside air pressure. The three typical VIPs samples are illustrated in Fig. 3 and the main parameters of core materials are demonstrated in Table 1.

The pre-treating procedure of the core material is essential and vital for the VIPs samples making, to remove the water and vapor embraced in the cavities (Di et al. 2013). However, the treating crafts are a little different for different core materials. After the pre-treating (in Fig. 4a), the core mate should be promptly carrier into the barrier envelopes prepared in advance and the vacuum heating seal should be taken in the special machine (in Fig. 4b).

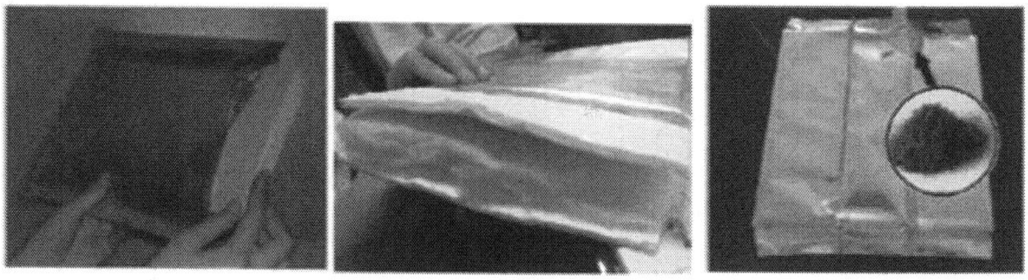
a b c

Figure. 3. The samples of VIPs with three typical porous core materials: (**a**) micro open cellular foam polyurethane; (**b**) superfine fibrous glass mate (**c**) nano-grain silica dioxides mixed with a few short glass fibers

The vacuuming, heat sealing, cooling and air charging proceeding are program controlled by the computer, and the setting air pressure is the final vacuum degree in the vacuum space. The setting rank of air pressure in these experiments are 0.1 Pa, 1 Pa, 10 Pa, 100 Pa and 1000 Pa. And then we got five VIPs samples with five different kind of inside air pressure. And the final vacuum inside the core materials are estimated by the anti- air pressure method (Di et al. 2013).

5.2 ETC testing and data collection

The ETC. of VIPs samples are strictly collected based on the Chinese standard GB/T3399-2009., and the ETC. measurements were conducted with a thermal conductivity meter (Fig. 5). The ETC. measurements in our laboratory can confirm values even lower than 1 mW/(m · K). As the thermal conductivity of barrier laminate is much bigger, even 1000 times, than that of the core materials under the vacuum condition, some heat may transfer from the edge of the VIPs but not through the panels, that is, thermal bridge effect (Tenpierik & Cauberg 2007). So the ECT collected by the equipment is a little deviation from the whole body's thermal conductivity. Here, we take the thermal conductivity of the center region as the ETC. of VIPs samples and the effect of thermal bridge is ignored. The data were collected in the Table 2.

Table 1. The main parameters of the core materials

	Parameters					
	Dimension mm × mm × mm	Density Kg/m^3	Pore diameter	Porosity	Mean diameter	Barrier membrane
open cellular foam polyurethane	300 × 300 × 20	50 ~ 75	80 ~ 100 μm	95 %	72 μm	High barrier laminates
superfine fibrous glass mate	300 × 300 × 20	200 ~ 250	2 ~ 10 μm	>90 %	6 μm	High barrier laminates
nano-grain silica dioxides mixed with a few short glass fibers.	300 × 300 × 20	180 ~ 200	10 ~ 40 nm	>95 %	32 nm	High barrier laminates and nylon film

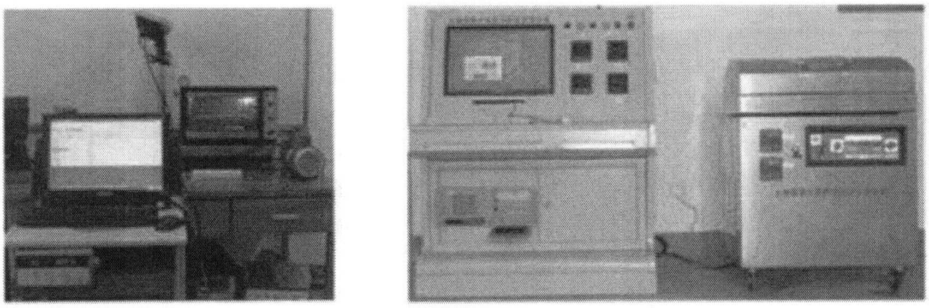

a b

Figure. 4. The VIPs making equipments: (**a**) The vacuuming and heating alternation pre-treating equipment; (**b**) program controlling VIPs making machine

Figure. 5. Laboratory measurement meter of thermal conductivity

Because the thickness of the barrier films (always several micro meters) is much smaller than that of core materials mate (in centimeter) and the thermal conductivity is much bigger than that of core materials under vacuum condition, the thermal resistance of the barrier films is ignored in the calculation and the further analysis. However, if the numerical model is used to predict the ETC. when the inside air pressure is near the barometric pressure, the thickness and thermal resistance should been taken into account.

Table 2. ETC. of VIPs samples and the inside air pressure in the measurements

	Core porous materials														
	open cellular foam polyurethane					superfine fibrous glass mate					nano-grain silica dioxides				
NO.	PU1	PU2	PU3	PU4	PU5	FG1	FG2	FG3	FG4	FG5	SI1	SI2	SI3	SI4	SI5
Setting air pressure/Pa	0.1	1	10	10^2	10^3	0.1	1	10	10^2	10^3	0.1	1	10	10^2	10^3
Final air pressure/Pa	5	13	45	140	713	0.6	7	25	165	763	15	38	80	241	892
Testing ETC./mW/(m·K)	4.12	4.15	5.04	16.23	21.16	2.68	2.96	3.19	5.84	14.63	3.06	3.82	4.14	4.75	8.24
Calculated ETC. mW/(m·K)	4.03	4.42	4.84	14.33	19.21	2.61	2.83	2.94	4.86	10.44	3.23	3.35	3.98	4.12	8.08

6. COMPARISONS AND DISCUSSIONS

The numerical prediction results and the testing results are compared in Fig. 6.

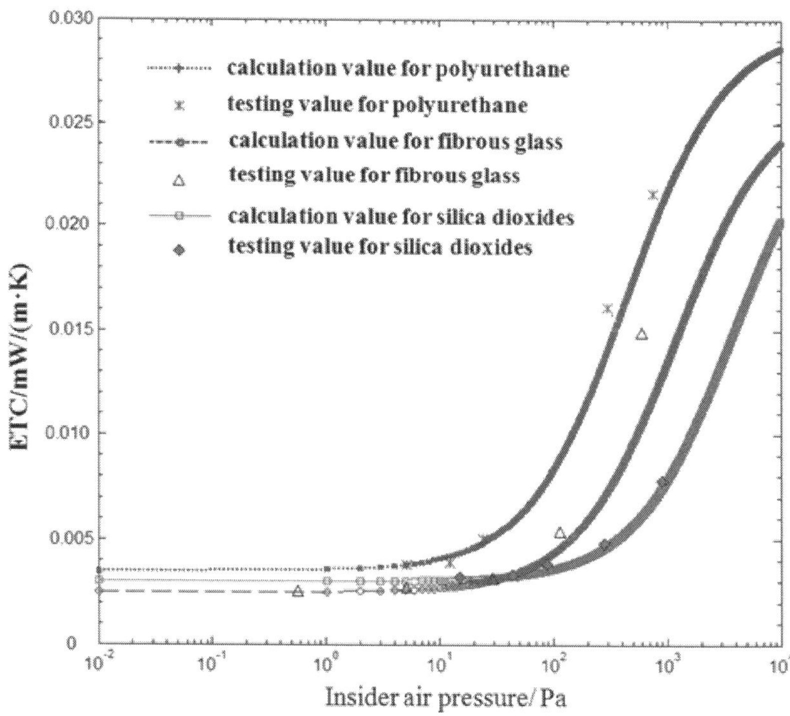

Figure. 6. The ETC comparison of the numerical calculations and experimental measurements for the three typical VIPs samples at different insider air conditions

When the inside air pressure is lower than 10 Pa, the ETC. curved line of the open cellular polyurethane is flat, basically around a constant value 4 mW/(m · K), and the theoretical calculation value is nearly the same, but slightly lower than the measured one. When the inside air pressure reaches for 100 Pa and continues to rise, the ETC., both of numerical prediction value and measurement one, sharply increased and over 11 mW/(m · K)(the limited value for so called adiabatic materials). The deviation between the theoretical calculations and the measured values gradually increases with the air pressure increases. The reason is that, the number of the air molecules in the pores increases and the frequent collision of the air molecules makes the free air path smaller than the cell feather size. The air thermal convection occurs and plays an important role by degrees. So the measured value is higher than that of the calculated one. So, the numerical ETC. prediction model for VIPs with the open micro cellular polyurethane, is much accurate when the inside air pressure is lower than 100 Pa.

For the superfine fibrous glass materials, the measured ETC. is agreeable with the calculated one when the inside air pressure is lower than10Pa and the ETC. value is around 3 mW/(m · K). As the inside air pressure exceeds 20 Pa and continues to rise, ETCs, both measured value and theoretical calculated one, sharply rise with the increasing of air pressure. And the difference between the two values also enlarges. One reason is that, the air molecules gathered in the pore leading to the thermal convection gradually intensified, as mentioned above. Another reason is that, the fibrous glass mate is flexible and compressible. The connected thermal resistance exists between two superfine fibrous pips and changes with the various air pressure differences between air pressure inside and outside the panels. The connected thermal resistance decreasing along with the inside air pressure increasing makes the measured ETC. value is slightly higher than that of calculated ones. So, the numerical ETC. prediction model for VIPs with the superfine fibrous glass core materials, is much accurate when the inside air pressure is lower than 10 Pa.

The ETCs of the nano grain silica dioxide powder as the core materials under the vacuum condition, both of the numerical prediction value and the measured one, are nearly the same and not more than 4 mW/(m · K) when the inside air pressure is no more than 100 Pa and theoretical calculation and measured values are in good agreement. The same trend is also displayed the similarities with the other two core materials. The reason is also mentioned above. The numerical ETC. prediction model for VIPs with the nano grain silica dioxide core materials is more accurate when the inside air pressure is lower than 100 Pa.

7. CONCLUSIONS

- The vacuum degree is one of the key external effect factors of the VIPs thermal property. Vacuum maintenance is an important and vital for the thermal stability and service life of VIPs. From the numerical simulation, the conclusions can be made that:

- ➢The feature size of micro open cellular polyurethane as the core materials for VIPs, should be within 100 μm and the inside air pressure should be not more than 1 Pa;

- ➢The feature size of superfine fibrous glass should be within 10 μm, and the inside air pressure should be not more than 10 Pa;

- ➢The feature size of nano- grain silica dioxide power should be within 100 nm, and the inside air pressure should be not more than 100 Pa.

- Compared the theoretical prediction value with the measured one, the agreement is very well when the inside air pressure is low. Even the ETC. same change trend, both of the numerical model and the measured values with increasing the air pressure, displaying in the research work, the deviation between the two gradually enlarged. So the numerical model can accurately predict the ECT of porous media under the low air pressure, especially under the P_{low}.

- Taking into consideration the properties of the core materials and vacuum degree, combined with the gas permeability rate of the barrier membrane, this numerical model can also be used to predict the service life of VIPs for further research, for the purpose of manufacture, engineering design and application.

Nomenclature

- d_g the diameter of the gas molecule, m
- ETC. effective thermal conductivity W/(m · K)
- K_B Boltzmann constant, 1.38×10^{-23} mJ/K
- K_n Knudsen number
- l_c the thickness of the plate core materials, m

- l_m gas molecule mean free path, m
- P Air pressure, Pa
- T temperature in thermodynamic scale, K

Greek symbols

- β universal constant, 1.5
- δ the mean diameter of the pore, m
- σ Steve Boltzmann constant, 5.6697×10^{-8} W/(K$^4 \cdot$ m^2)
- φ the attenuation coefficient,445 m^{-1}
- ξ the porosity of the core materials
- λ thermal conductivity W/(m · K)

Subscripts

- c thermal conduction
- e effective
- g gas
- r thermal radiation
- s solid phase

ACKNOWLEDGEMENTS

This research is financially supported by the Natural Sciences Foundation of Shanghai, China (15ZR1419900). The authors also sincerely acknowledge the technical support from Prof. Zhang Xuelai for his valuable suggestion and contributions in mounting the experimental set-up and installing the data acquisition devices.

AUTHORS' CONTRIBUTIONS

AK and LK carried out the experimental and theoretical studies, and drafted the manuscript. CW and DC participated in the sequence alignment and drafted the manuscript. All authors read and approved the final manuscript.

REFERENCES

1. Alam M, Singh H, Limbachiya MC (2011) Vacuum Insulation Panels (VIPs) for building construction industry–A review of the contemporary and future directions. Appl Energy 88:3592–3602
2. Baetens R, Jelle BP, Thue JV et al (2010) . Vacuum insulation panels for building application: a review and beyond. Energy Buildings 42:147–172
3. Bouquerel M, Duforestel T, Baillis D, Rusaouen G (2012a) Heat transfer modeling in vacuum insulation panels containing nano-porous silicas: A review. Energy Buildings 54:320–336
4. Bouquerel M, Duforestel T, Baillis D, Rusaouenb G (2012b) Mass transfer modeling in gas barrier envelopes for vacuum insulation panels: a review. Energy Buildings 55:903–920
5. Brunner S, Stahl T, Ghazi Wakili K (2012) An example of deteriorated vacuum insulation panels in a building facade. Energy Buildings 54:278–282
6. Caps R, Fricke J (2000) Thermal conductivity of opacified powder filler materials for vacuum insulations. Int J Thermophys 21:2 445–2452
7. Caps R, Heinemann U, Ehrmanntraut M, Fricke J (2001) Evacuated insulation panels filled with pyrogenic silica powders: properties and applications. High Temp-High Press 33:151–156

8. Di X, Gao Y, Bao C, Hu Y (2013) Optimization of glass fiber based core materials for vacuum insulation panels with laminated aluminum foils as envelopes. Vacuum 97:55–59

9. Di X, Gao Y, Bao C, Ma S (2014) Thermal insulation property and service life of vacuum insulationpanels with glass fiber chopped strand as core materials. Energy Buildings 73:176–183

10. Fricke J, Heinemann U, Ebert HP (2008) Vacuum insulation panels-From research to market. Vacuum 82:680–690

11. Jung H, Jang CH, Yeo IS, Song TH (2013) Investigation of gas permeation through Al-metalized film for vacuum insulation panels. Int J Heat Mass Transf 56:463–446

12. Kan A, Han H (2013) Effective Thermal Conductivity of Open Cell Polyurethane Foam Based on the Fractal Theory. Adv Mat Sci Eng Article ID 125267:7

13. Kim J, Song TH (2013) Vacuum insulation properties of glass wool and opacified fumed silica under variable pressing load and vacuum level. Int J Heat Mass Transf 64:783–791

14. Kim J, Jang C, Song TH (2012) Combined heat transfer in multi-layered radiation shields for vacuum insulation panels: Theoretical/numerical analyses and experiment. Appl Energy 94:295–302

15. Kwon J-S, Jang CH, Jung H, Song TH (2009) Effective thermal conductivity of various filling materials for vacuum insulation panels. Int J Heat Mass Transf 52:5525–5532

16. Mendes MAA, Ray S, Trimis D (2013) A simple and effective method for the evaluation of effective thermal conductivity of open-cell foam-like structures. Int J Heat Mass Transf 66:412–422

17. Mendes MAA, Talukdar P, Ray S, Trimis D (2014) Detailed and simplified models for evaluation of effective thermal conductivity of open-cell porous foams at high temperatures in presence of thermal radiation. Int J Heat Mass Transf 68:612–624

18. Nussbaumer T, Bundi R, Tanner C, Muehlebach H (2005) Thermal analysis of a wooden door system with integrated vacuum insulation panels. Energy Buildings 37:1107–1113

19. Nussbaumer T, Ghazi Wakili K, Tanner C (2006) Experimental and numerical investigation of the thermal performance of a protected vacuum insulation system applied to a concrete wall. Appl Energy 83:841–855

20. Schwab H, Heinemann U, Wachtel J (2005a) Prediction for the increase in Pressure and water content of vacuum insulation panels (VIPs) Integrated into Building Construction using Model Calculation. J Therm Envel Build Sci 28:327–344

21. Schwab H, Heinemann U, Beck A (2005b) Prediction of service life for vacuum insulation panels with fumed silica kernel and foil cover. J Therm Envel Build Sci 28:357–374

22. Tenpierik M, Cauberg H (2007) Analytical models for calculating thermal bridge effects caused by thin high barrier envelopes around vacuum insulation panels. J Build Phys 30:185–215

23. Wang M, Pan N (2008a) Modeling and prediction of the effective thermal conductivity of random open-cell porous foams. Int J Heat Mass Transf 51:1325–1331

24. Wang M, Pan N (2008b) Predictions of effective physical properties of complex multiphase materials. Mater Sci Eng R 63:1–30

25. Wang M, Pan N, Wang J, Chen S (2007) Mesoscopic simulations of phase distribution effects on the effective thermal conductivity of microgranular porous media. J Colloid Interface Sci 311:562–570

Chapter 5

Numerical and experimental thermal analysis for the improvement of various types of windows frames and rolling-shutter boxes

Nicola Cardinale, Gianluca Rospi , Tiziana Cardinale

Department of European and Mediterranean Cultures: Architecture, Environment, Cultural Heritages (DICEM), University of Basilicata

ABSTRACT

Data provided by Italian Ministry of Environment say that in 5 years you consume, just for heating of an apartment, an amount of energy equal to that required for the all construction of the same apartment. The years come down to three if one includes other energy consumption. In terms of primary energy, the building-plant system is responsible for about 45 % of the national energy demand. The rolling-shutter box system, in most cases, is the first responsible for thermal dispersions and input of free heat coming from outside. In addition, the window frames have the obligation of the CE marking to be sold on the free European market. Among the basic requirements, the thermal transmittance is the parameter responsible for the energy performance. The research concerned the energy study of different window frames and rolling-shutter boxes through the use of finite element methodology and experiments in situ. This study was carried out on six different types of windows frames (PVC, wood, and wood-aluminium) and four different models of rolling shutters-boxes (all with structure made by expanded polystyrene). The experimental analysis allowed the comparison between the thermal transmittance values calculated according to the UNI EN ISO 10077-1/2 and measured in accordance with ISO 9869.

KEYWORDS

Experimental and thermal analysis, Thermal transmittance, Energy efficiency of windows frames and rolling-shutter boxes, Finite element analysis

1. INTRODUCTION

Improving the energy efficiency of the building and the indoor comfort is just some of the goals that the European Union aims to achieve by 2020. In this sense, the EU has promoted various programs, projects and guidelines, such as 2002/91/EC, 2006/32/EC, and 2010/31/EU, which define addresses, tools, policies, and solutions in terms of energy efficiency of new and existing buildings.

"The EC marking indicates that the product complies with all the Community provisions which provide its use", from design to manufacture, placing on the market, putting into service of the product until its disposal. The EC marking governs the entire life cycle of the product from the time of placing on the market and it is governed by Directive 98/34/EC. With regard to the performance characteristics, the Directive 98/34/EC refers to harmonized European standards.

For the fixtures, the harmonized European standard is UNI EN 14351-1 [product standard] that specifies the performance characteristics of windows and external pedestrian doors and includes the framework of performance requirements, and test criteria necessary to ensure conformity with the essential requirements set by Directive 98/34/EC. It should be noted that, the civil sector, in Italy, representing over 40 % of the national energy demand and recorded a progressive and inexorable growth of consumption. An Italian building expresses very low energy performance recording annual consumption, in terms of end use, that vary from 160 kWh/m^2 per year to more than 230 kWh/m^2 per year, compared with lower consumption between 30 and 60 % within EU level. A so high energy consumption determines, as a result, values of annual emission of greenhouse gases well above the European average (19 million of installed boilers in Italy consume more than 25 Mtoe per year determining an emission of about 80 Mtonn of equivalent CO_2 per year).

In the past couple of decades, the glass has been increasingly used in the architectures; and today the windows are a key material in façade technology. When employing windows in modern buildings, numerous physical aspects need to be considered such as visual contact between interior and exterior, use of daylight, optimizing and controlling solar energy gains, minimizing thermal losses, optimizing thermal comfort, noise protection, air and driving rain tightness, and fire safety. The area of the windows in the building envelope is approximately the 20 % of the total area envelope and the overall heat losses through the windows is even more than four times higher than for the insulated building walls. One of the old methods of suppressing the heat losses through windows is the use of wooden shutter closed by night. Today, we use mostly the various types of blinds and among them e.g., the roller blinds mainly to provide shade, privacy and as well as the security (usually external roller blinds).

With regard to experimental studies in the field of windows, we focused on the following:

- An experimental study to determine the overall heat transfer coefficient of the double-glazed window without low-emission coating and the heat losses through window glazing that use roller blinds was published in [1, 2].

- In [3], we find a study on window shutters that using PCM material for reduce the heat gain by absorbing the heat gain before it reaches the indoor space. Instead in [4], the application results of a previous and current EU-project on super-insulating glazing based on monolithic silica aerogel were described.

- Some research on the effects of the geometric characteristics of the cavities of the aluminum frames and the emissivity properties of the cavity inner surfaces was published in [5].

- In [6, 7], the authors study the effect of the thermal transmittance and solar transmittance of some windows on the building thermal balance. In these studies, different climatic localities were considered. Instead in [8], the energy savings connected to the potential use of the reversible windows in several Italian locations for buildings with no-prevalent cooling or heating requirements were analyzed.

- In [9], we find a simple method for analyzing and discussing energy flows through glazing. This method is based on monthly mean values of interior-exterior temperature difference and solar irradiance, and it was implemented for eight case study locations in Europe namely Bucharest, London, Madrid, Moscow, Rome, Stockholm, Warsaw, and Zurich.

- In [10], the focus is on buildings in Greece describing the impact of glazing on energy consumption and comfort.

- In [11], the authors dealt with the role of geometry factors, such as window orientation, window to wall ratio, and room width to depth ratio, on building energy performance in a commercial office building. The analysis was developed on office buildings located in various climate zones.

- In [12], two different two-dimensional frame/spacer heat transfer calculation methodologies used in North America and in Europe were compared. The [12] shows algebraically that the differences between the ISO and ASHRAE methods turn out to be due to the way the corner regions of the window frame and glazing are treated for the evaluation of the thermal transmittance by two-dimensional calculations.

2. ANALYZED WINDOW FRAME-ROLLING SHUTTER BOX SYSTEM

The system window frame and rolling-shutter box are on the border between outdoor and indoor space; it represents the view to the outside and the entry of natural light in indoor spaces and a point of discontinuity and of heat loss; so a good design provides an excellent protection of indoor environments from the elements of external environment.

According to the requirements of the municipal and sanitary building regulations, one of the necessary requirements for the habitability of indoor environments is that the window or French window of a room must provide at least an opening of 1/8 to the surface of the floor of the room itself.

In this work, different types of window frame and rolling-shutter box systems are analyzed. Specifically four different types of frames were studied, characterized by constructive material (PVC, wood, and, wood-aluminum) and four different types of rolling-shutter box made by EPS, characterized by construction type (different thicknesses of the polystyrene and different positions with respect to the window frame).

3. ANALYZED FRAMES

We analyzed two open joint typologies of PVC frames with frame with five air spaces of thickness 70 mm and shutter with seven air spaces of thickness from 70 to 80 mm. The air space profiles, suitable to contain closed or C-shaped metal reinforcements of different thicknesses, is separated from the outside, so it does not ever come in contact with water or other atmospheric agents. The standard glass mounted on the windows is 4/20/4 (Fig 1a).

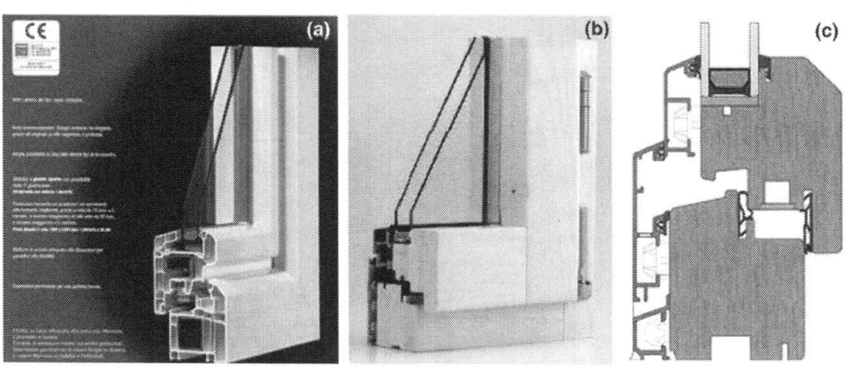

Figure. 1. Analyzed frame typologies **a.** PVC, **b.** Wood, **c.** Wood-Aluminum

As regards, instead, the wooden frame analyzed, this is characterized by the construction typology by laminated wood of section 68 × 78 mm both for the shutter and for the frame. The air between shutter and frame is 12 mm, while the beating is 18 mm. The wood is joined by means of tenons at 90°. The glass air space used is 4/20/4 mm (Fig 1b).

Finally, the wood-aluminum frame typology is characterized by the combination of two materials, laminated wood with section 68 × 78 mm for shutter and frame, coated on the outside with an anodized aluminum profile pegged to the wood frame. The glass air space used is 4/20/4 mm (Fig 1c).

While as regards to the rolling-shutter boxes, all four models are made by expanded polystyrene, said technically EPS. The EPS has a closed-cell structure composed of 98 % air, which in addition to thermal insulation gives to this product property of vapor barrier.

In the past, the traditional rolling-shutter boxes represented weak points that often allowed leaks of cold and humidity. In the analyzed rolling-shutter boxes, these issues are resolved in a definitive way, through the use of polystyrene. The natural and continuous ventilation in the rolling shutter and the perfect closure between the box and the frame, prevent the formation of mold and condensation both in the box and inside the rooms. Among the four models analyzed, there are two with inspection by the front, called Simply Figs. 2 and 3 and two with inspection by the bottom, called Hyde Figs. 4 and 5.

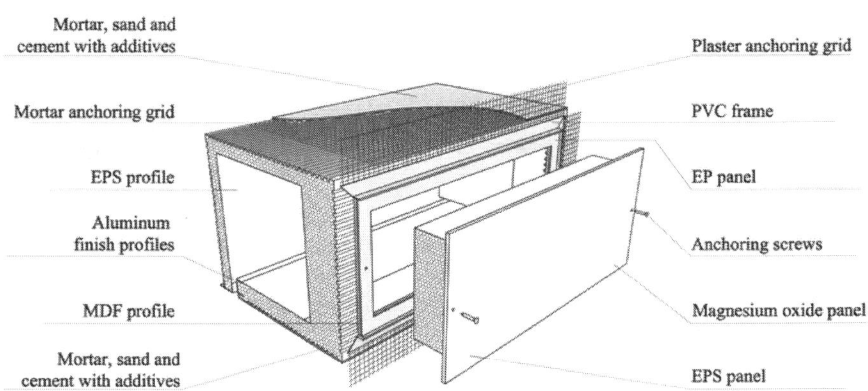

Figure. 2. Simply model assonometry

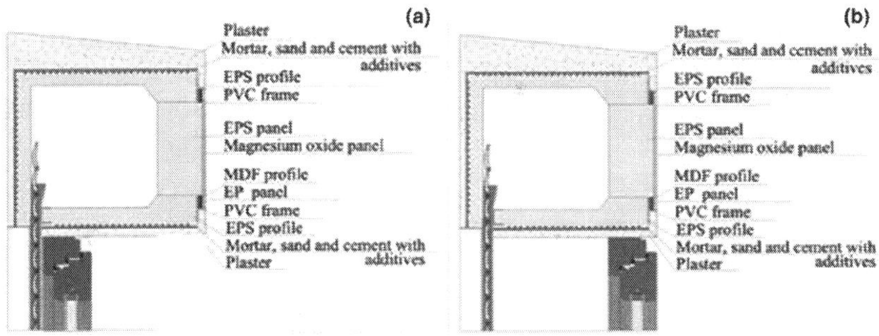

Figure. 3. Positioning window frame on rolling-shutter box: **a** outer edge, **b** inner edge

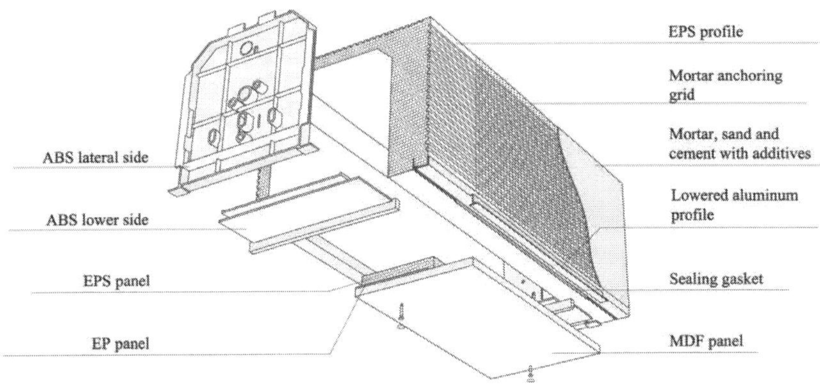

Figure. 4. Hyde model assonometry

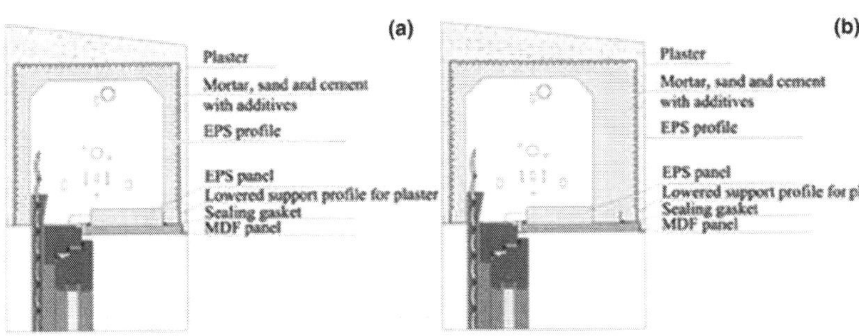

Figure. 5. **a** Hyde rolling-shutter box 30-thickness 30 mm, **b** Hyde rolling-shutter box 35-thickness 35 mm

The rolling-shutter boxes with front inspection Simply analyzed are characterized by having a structure in self-extinguishing expanded polystyrene of 30 kg/m^3 and a hatch composed of a removable panel in expanded polystyrene glued by polyurethane glues on a panel of magnesium oxide treatable with paints of water. Specifically, the two models were analyzed Simply 30 (characterized by having a thickness of 30 mm of EPS) and the model Simply 35 (characterized by having a thickness of 35 mm of EPS) both analyzed in two different configurations of mounting frame (outer edge wall and inner edge wall) (Fig. 3).

The rolling-shutter boxes with inspection by the bottom, called Hyde, are characterized by having a structure in flame retardant expandable polystyrene of 25 kg/m^3, covered externally with an anchoring network and mortar, sand, and cement admixed with clinging latex, to give maximum set to the application of plaster on the walls. The flap plaster lowered by 2 cm compared to the profile of the box, allows you to completely hide the front side of the celino, while also serving as support by the inclusion of the same. The sealing gasket in contact with the surface of celino offers improved air tightness and optimal insulation performance. The particular morphology of this type of celino, offers the possibility to thicken considerably the lower insulation, and in some cases, lead to a thickness of the panel in EPS equal to 4 cm. Specifically, two models were analyzed: Hyde 30 (characterized by having a thickness of 30 mm of EPS), and the model Hyde 35 (characterized by having a thickness of 35 mm of EPS) (Fig. 5).

4. METHODOLOGY OF STUDY

As previously mentioned the norm, UNI EN 14351-1 Part 1 is the Italian normative reference for the EC marking of transparent windows. The legislation specifies that the value of the thermal transmittance of the window including frame can be obtained either by laboratory testing or by simple calculation according to EN ISO 10077-1, or through finite element calculation according to UNI EN ISO 10077-2 in combination with the simplified calculation.

For the study of rolling-shutter box-frame systems, it was adopted a research methodology complemented by finite element analysis (FEM) and experimental measures on-site. Through the campaign of experimental measurements, it was possible to schematize the numerical model and calibrated FEM.

The structure of the research work can be divided into three main phases:

1 an experimental phase that saw the conductance measurement on-site through the methodology described by the standard prEN 15203;

2 validation and calibration of the numerical model FEM schematize for the frame system and of the rolling-shutter box system;

3 extension of the FEM analysis for different types of window frames and boxes and calculation of thermal transmittance according to the UNI EN ISO 10077-2;

4 evaluation of improvement hypothesis on the rolling-shutter box-frame systems analyzed.

The program used for the schematization of analyzed systems, the FEM analysis and the verification of thermal transmittance is a commercial software: the FRAME SIMULATOR 2 Dartwin in the Student Version.

4.1 Experimentation

The in situ campaign developed at the University of Basilicata, Matera, was to measure the surface temperatures and heat flux on the specific window frame system made in PVC and on the rolling-shutter box system Simply 30. On-site analysis was done by mounting the system on a partition wall between two environments, which were air conditioned in order to create constant thermal difference (Fig. 6).

Figure. 6. Installation of the system and sensors for testing on-site

After adjusting appropriately the climatic parameters of the two rooms in order to achieve a medium Δt of about 7 °C, we mounted the frame,

This condition guaranteed, in fact, of detecting the magnitudes (species thermal flows) in the conditions in which they assume higher values. In this way, it was possible to operate with high output signals from the transducers, obtaining as a result that: the system is less influenced by external disturbances, the measuring instruments operate in ranges significantly greater than the minimum threshold of detection and the changes of the signals are significantly higher than the minimum resolution of the instruments.

Subsequently, through the aid of a multi data logger, four surface temperature sensors and a thermal flow meter, it was initiated the measurement campaign for a period of about 20 days, making a total of 1130 measurements.

The acquired data were processed on-site later with a typical method of systems analysis in transient conditions: the progressive average method.

In the method called "progressive average" or "moving average" one can calculate wall conductance, from the sampled values of surface temperature and heat flux with the following equation:

$$C = \frac{\sum_{f=1}^{N} q_j}{\sum_{f=1}^{N} \left(T_{\mathrm{Sif}} - T_{\mathrm{SFj}}\right)} \frac{\mathrm{W}}{\mathrm{m^2K}}$$

As N increases, the ratio tends to converge to a stationary value not influenced by the thermal mass of the wall. This can easily be seen from the graph that shows the different values obtained from the calculations; in fact, the function converges with oscillations around a horizontal asymptote with a maximum amplitude of about 0.05 W/m^2 K.

Through experimental phase, it was possible to validate and calibrate the numerical model FEM and to calculate the error.

The results of measurements on the PVC window frame system were summarized in the following Tables1and 2. The first shows the average measured data and the second one describes the difference between the measured conductance in situ and that calculated according to the UNI EN ISO 10077-2 through the application of the FEM method:

Table 1. Average values of the thermophysical measurement on the frame

Average temp. of heated room (°C)	Average temp. of conditioned room (°C)	Average heat flux Φ (W/m^2)	Conductance (progressive average method) (W/m^2 K)	Transmittance (normed surface resistance normed 0.13 m^{2}K/W) (W/m^2 K)
27.05	20.35	9.6	1.689	1.173

Table 2. Measured and calculated thermal transmittance with relative percentage error

Measured thermal transmittance (W/m^2 K)	Calculated thermal transmittance (W/m^2 K)	Error (%)
1.173	1.122	4.55

As regards, the experimentation on the rolling-shutter box system Simply 35, the measurements returned a conductance value equal to 0.870 W/m^2 K compared to a calculated value of 0.835 W/m^2 K.

4.2 Thermal performances of discussed systems

After having validated the software, it was possible to take the next step of analysis of the thermal transmittance of other types of window frames and similar rolling-shutter boxes. In total, the analyzed systems were eight; of which three windows, two with PVC frame, one with wood frame and one with wood-aluminum frame and four boxes for rolling shutters all in EPS, but with different construction technologies.

4.3 Evaluation of window frame system

As regards window frames, once built the numerical model on the software, temperatures and heat flows trends were analyzed and then calculated the value of the thermal transmittance in accordance with the UNI EN ISO 10077-2.

Below there is the schematization of the frame system into the FEM software with the indication of the relative materials placed Fig. 7.

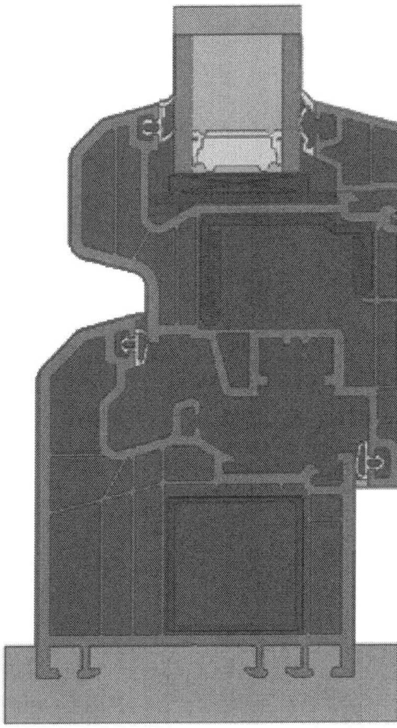

Fig.ure 7. Schematization of PVC window frame

Materials list in Table 3.

Table 3. Materials list

Name	λx [W/mK]	λy [W/mK]	ε	Color
EPDM	0.250	0.250	0.900	
Stainless steel	17.000	17.000	0.900	
Air	0.106	0.115	0.900	
PVC	0.170	0.170	0.900	
Common glass	1.000	1.000	0.900	
Adiabatic	0.000	0.000	0.900	
Air	0.028	0.027	0.900	

Table 4. Boundary conditions of the FEM model

	External boundary conditions	Internal boundary conditions
Temperature	0.00 °C	20.00 °C
Surface resistance	0.04 m^2K/W	0.13 m^2K/W

As regards the internal and external hypothesized boundary conditions, these are summarized in the table below.

Through the analysis of the temperature and direction of heat flow was possible to study in detail, the structural problems of the frame and joints between fixed and mobile frame, often points of heat loss (Fig. 8).

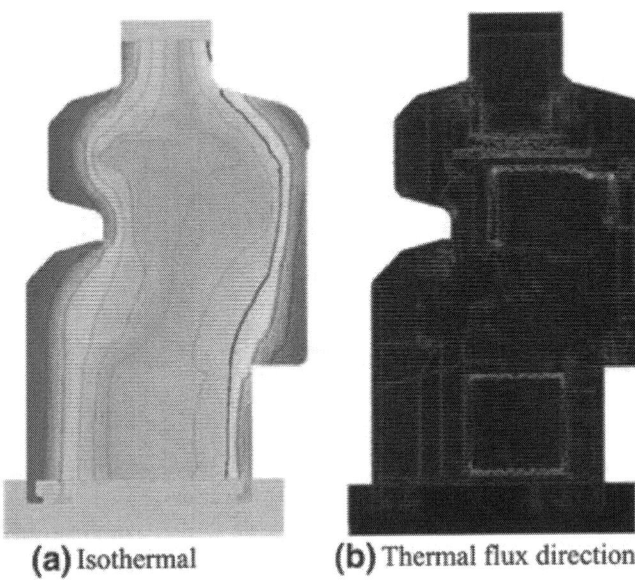

Figure. 8. Values of temperature and heat flux in the PVC frame

The calculation of the thermal transmittance, according to the UNI EN ISO 10077-2:2012, returned a value of the thermal transmittance of the only frame Uf equal to 1.122 W/m^2 K and the thermal transmittance of the frame more glass Uw equal to 1.847 W/m^2 K.

The same procedure was done on other models of frames returning the values summarized in Table 5.

Table 5. Thermal transmittance value

Model	Thermal transmittance frame uf (W/m^2 K)	Thermal transmittance frame/glass uw (W/m^2 K)
PVC 70 mm shutter	1.572	1.852
PVC 80 mm shutter	1.122	1.847
Wood 68 mm shutter	1.678	2.272
Wood–aluminim 58 mm shutter	1.618	2.053

4.4 Evaluation of rolling-shutter box system

Regarding the rolling-shutter boxes, also in this case, after the construction of the numerical model on software, the next step was to analyze the trend of temperatures and heat flows and then calculate the value of the thermal transmittance in accordance with the UNI EN ISO 10077-2.

For the rolling-shutter boxes, the analysis was made with the same constructive material (all are made by EPS), the only changes relate to the different conformations of construction, the different thicknesses of EPS, and the different dispositions of the frame.

For example with regard to the rolling-shutter box system called Simply 35 the analyzes were two, one with the mounting of the frame in correspondence with the external edge of the box and one in correspondence of the inner edge of the box (Fig. 3).

As for the window frame system also for the rolling-shutter box system, the first operation was to schematize the FEM software (Fig. 9).

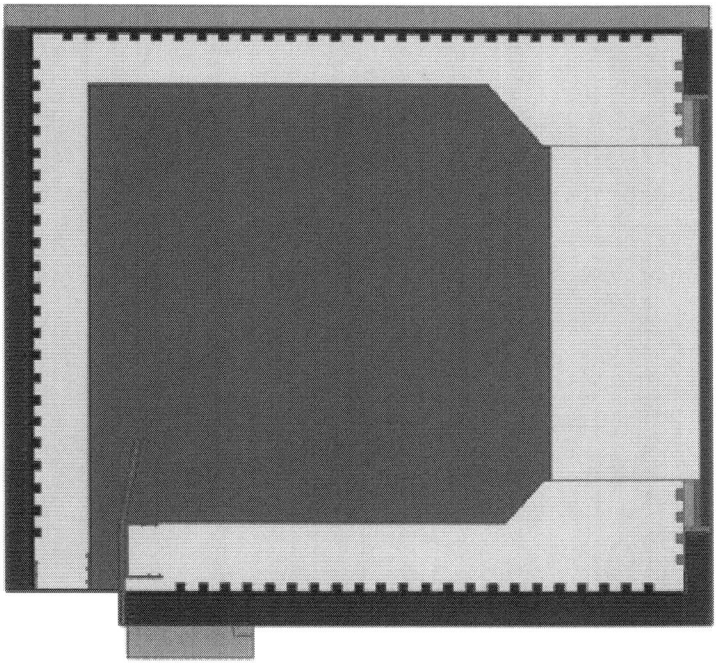

Figure. 9. Schematization of rolling-shutter box with frontal inspection Simply 35

The boundary conditions were the same as in the Table 4.

The next step was to analyze the trend of temperatures and heat flux to obtain then the value of the thermal transmittance according to the UNI EN ISO 10077-2. In Fig. 10, the different configurations of the isotherms are shown in the two different cases of the frame installation.

The calculation of the thermal transmittance, according to the UNI EN ISO 10077-2:2012, returned a value of thermal transmittance for the Simply 35 system outer edge of Uf equal to 0.835 W/m^2 K and a thermal transmittance value of 0.573 W/m^2 K for the Simply system inner edge.

The same procedure was done on other models of boxes and returns the values summarized in Table 7.

Materials list in Table 6.

Table 6. Materials list

Name	λx [W/mK]	λy [W/mK]	ε	Color
Adiabatic	0.000	0.000	0.900	
EPS Polystyrene closed-cell foam	0.034	0.034	0.900	
Plasters and mortars	0.700	0.700	0.930	
Mortar of sand and additivated cement	0.800	0.800	0.930	
Unpainted aluminum	160.000	160.000	0.300	
PVC-U rigid	0.170	0.170	0.900	
Mdf panel	0.150	0.150	0.930	
Pe mat	0.040	0.040	0.930	
Magnesium oxide panel	0.216	0.216	0.930	

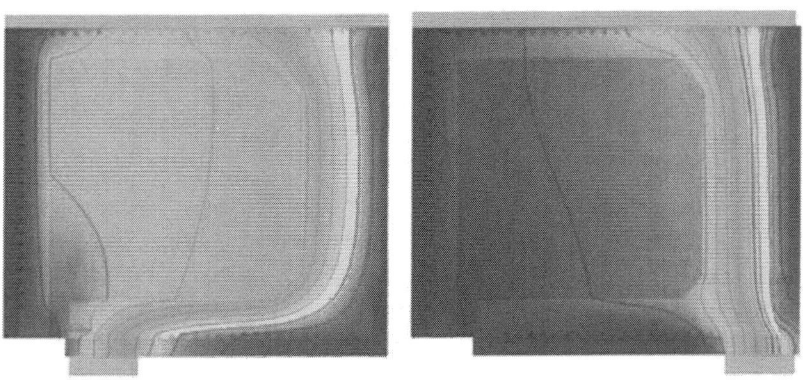

(a) Isothermal Simply 35 shutter external (b) Isothermal Simply 35 shutter internal

Figure. 10. Isoterme Simply 35 with configuration frame outer edge (**a**) and frame inner edge (**b**)

Table 7. Thermal transmittance value

Model	Thermal transmittance Uf inner edge (W/m^2 K)	Thermal transmittance Uf outer edge (W/m^2 K)
Simply 30 EPS 30 mm	0.838	0.945
Simply 35 EPS 35 mm	0.576	0.835
Hide 30 EPS 30 mm	Not supported	1.038
Hide 35 EPS 35 mm	Not supported	0.703

4.5 Analysis of improvements

The analysis of the improvements was done considering separately the window frame system and the rolling-shutter box system. As regards the window frame system, the improvement was hypothesized for the model of PVC shutter with thickness equal to 80 mm and for the wooden model. The improvement hypothesis was those of replacement of double glazing with a low-emissivity one and replacement of an air interspace with krypton gas. These two improvements allowed to intervene both on emissivity bringing the value of ε and from 0.9 to 0.02 and on the conductivity of the interspace halving the value ($\lambda = 0.01$ W/m K). The final results on the window frame system are summarized in the Table 8.

As regards instead the rolling-shutter box system, the hypothesized improvement was to insert inside the rolling-shutter box a thermal reflective insulation composed by a double sheet of aluminum with an interposed inert air bubble made of polyethylene with high density of total thickness equal to 6.5 mm (Fig. 11).

Table 8. Thermal transmittance value before and after

Model	Transmittance values Uw		
	Before (W/m^2 K)	After (W/m^2 K)	
PVC 80 mm shutter	1.847	1.664	Reduction 10 %
Wood 68 mm shutter	2.272	2.000	Reduction 12 %

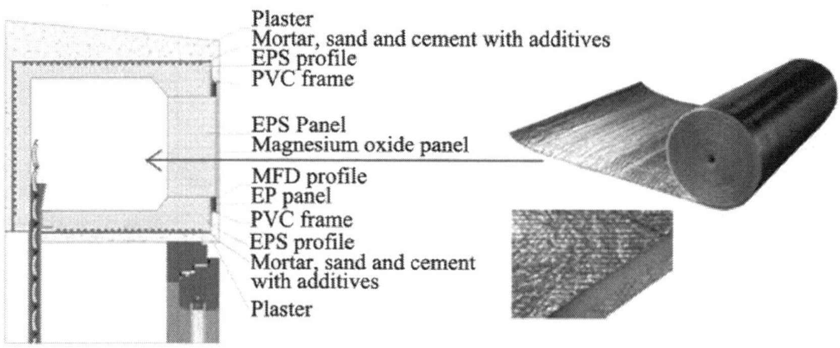

Figure. 11. Hypothesized thermal reflective insulation

The final results on the rolling-shutter box system are summarized in the Table 9.

Table 9. Thermal transmittance value before and after

Model	Transmittance values Uf		
	Before(W/m^2 K)	After(W/m^2 K)	
Simply 35 inner edge	0.576	0.400	Reduction 30 %
Simply 30 inner edge	0.945	0.638	Reduction 32 %
Hide 35 outer edge	0.703	0.436	Reduction 37 %
Hide 30 outer edge	1.038	0.701	Reduction 31 %

5. CONCLUSIONS

The commercial finite element method presented in this work, validated by our experimental analysis, is able to appraise, with good accuracy, improvements in thermal performance to be applied to windows and rolling-shutter box systems. Finally, we want to highlight the significant reduction of the thermal transmittance, at least 30 %, in the rolling-shutter box systems with the application of a thermal reflective insulation. This last is an innovative material designed to insulate the walls, but that, as demonstrated by the results obtained in this work, can be effectively used even in the insulation of rolling-shutter box systems.

REFERENCES

1. Oleskowicz-Popiel, C., Sobczak, M.: Effect of the roller blinds on heat losses through a double-glazingwindow during heating season in Central Europe. Energ. Build. **73**, 48–58 (2014)

2. Shahid, H., Naylor, D.: Energy performance assessment of a window with a hor-izontal Venetian blind. Energ. Build. **37**, 836–843 (2005)

3. Esam Alawadhi, M.: Using phase change materials in window shutter to reduce the solar heat gain. Energ. Build. **47**, 421–429 (2012)

4. Schultz, J.M., Jensen, K.I., Kristiansen, F.H.: Super insulating aerogel glazing. Sol. Energ. Mater. Sol. Cells. **89**, 275–285 (2005)

5. Asdrubali, F., Baldinelli, G., Bianchi, F.: Influence of cavities geometric and emissivity properties on the overall thermal performance of aluminum frames for windows. Energy. Build. **60**, 298–309 (2013)

6. Gasparella, A., Pernigotto, G., Cappelletti, F., Romagnoni, P., Baggio, P.: Analysis and modelling of window and glazing systems energy performance for a well-insulated residential building. Energ. Build. **43**, 1030–1037 (2011)

7. Seunghwan Yoo, et al.: Thermal transmittance of window systems and effects on building heating energy use and energy efficiency ratings in South Korea. Energ. Build. **67**, 236–244 (2013)

8. Gugliermetti, F., Bisegna, F.: Saving energy in residential buildings: the use of fully reversible windows. Energy **32**, 1235–1247 (2007)

9. Manz, H., Menti, U.: Energy performance of glazings in European climates. Renew. Energ. **37**, 226–232 (2012)

10. Stegou-Sagia, A., et al.: The impact of glazing on energy consumption and comfort. Energ. Convers. Manag. **48**, 2844–2852 (2007)

11. Irina Susorova, et al.: The effect of geometry factors on fenestration energy performance and energy savings in office buildings. Energ. Build. **57**, 6–13 (2013)

12. Petar Blanusa, et al.: Comparison between ASHRAE and ISO thermal transmittance calculation methods. Energ. Build. **39**(3), 374–384 (2007)

Chapter 6

An Assessment of Thermal Comfort in Multi Storey Office Buildings in Ghana

Barbara Simons[1*], Christian Koranteng[2], Emmanuel Adinyira[1], Joshua Ayarkwa[1]

[1]Department of Building Technology, Kwame Nkrumah University of Science and Technology, Kumasi, Ghana

[2]Department of Architecture, Kwame Nkrumah University of Science and Technology, Kumasi, Ghana

ABSTRACT

Amidst the recent development in the usage of curtain walls for office buildings, high utilization of energy and poor thermal comfort issues have become paramount. This paper assesses thermal comfort in multi storey (naturally and mechanically ventilated) office buildings in Accra, the Capital city of Ghana using Fanger's Predicted Mean Votes (PMV) and Predicted Percentage of Dissatisfied persons (PPD) model. The model relates to the American Society of Heating, Refrigerating and Air-Conditioning Engineers (ASHRAE) Standard 55 (which gives an acceptable temperature range of 23˚C - 26˚C) and the International Standards Organization (ISO) 7730. Indoor environmental parameters (temperature and relative humidity) of 4 multi storey office buildings were recorded over a 10 month period. The environmental parameters were analyzed using PMVcalc_v2 software which resulted in the generation of PMV-PPD values. The findings reveal high PMV-PPD values in the Naturally Ventilated Building (NVB) whiles the

Mechanically Ventilated Buildings (MVB) fall within the comfort zone. Meanwhile, the Actual Mean Votes (AMV) by the occupants suggest all four buildings are relatively comfortable with the mechanically ventilated offices being more comfortable. Additionally, it is recommended that buildings are orientated with their longer sides facing north-south, with enough shading in order to improve the thermal comfort conditions of work spaces.

KEYWORDS

Thermal Comfort; ASHRAE Standard 55; Office Buildings; Natural; Mechanical

1. INTRODUCTION

Man has over the years been successful in creating environment that is conducive for all kinds of activities. One of the basic requirements of this environment is to maintain thermal conditions comfortable for occupants as these have a direct effect on health, productivity and morale [1] . What is more, [1] ascertains that thermal comfort is attained when a thermal balance is achieved: a situation in which no heat storage occurs in the body. Although this can be achieved over a wide range of thermal environmental conditions, thermal-comfort is associated with conditions to which the body can readily adjust. Reference [2] defines thermal comfort as the "state of the mind that expresses satisfaction with the surrounding environment". Standards such as ASHRAE Standard 55 [2] and ISO Standard 7730 [3] have been the basis in the determination of thermal comfort by most practitioners and researchers [4] . Nonetheless, people in different climatic zones feel comfortable at different indoor air temperatures: a situation which can vary considerably from that of the world standards.

The factors affecting thermal comfort depend on four environmental (i.e. dry bulb temperature, mean radiant temperature, relative humidity and air velocity) and two personal (clothing-insulation and physical activity) parameters [5] . The thermal comfort standard prescribed by ISO 7730 was the first to have been used on a worldwide basis [6] . It's based on Fanger's work in climate chamber experiments using young Danish students and eventually the PMV model.

Architecturally, the hot and humid regions are the hardest climates to improve through design: due to high outdoor thermal conditions (humidity and temperature) that result in soaring indoor temperatures exceeding the ASHRAE summertime comfort upper limit of 26°C. A temperature limit should at least get 90% of occupants feeling thermally comfortable for most part of the year [7] .

1.1. Problem Statement

In recent years, most multi storey office buildings in Ghana have become what is termed the "glass box" exposing all or most part of its facades to the high day time temperatures. The effect of which is the contribution of thermal discomfort within the indoor spaces. In an attempt to ameliorate the effect of this exposure, multi storey office buildings have resorted to the use of complex HVAC systems which comes with its problems such as high usage of energy and greater emission of Carbon dioxide (CO_2) into the atmosphere.

1.2. Objectives of the Paper

The objectives of the research include the followings:

- To measure indoor and outdoor thermal comfort parameters of the selected buildings;
- To compare the predicted mean votes (PMV) within the office with the actual mean votes (AMV) given by the occupants.

2. LITERATURE REVIEW

2.1 The PMV Model

The Fanger-based approach, which is alternatively known as the heat-balance approach, is built on experiments in climate chambers and eventually theorized into the PMV-PPD model. It was fist propounded by

Fanger around the 1970s with about 1296 participants dressed in standardized clothing and completed standardized activities constituting a metabolism rate of 1.2 Mets or 70 W/m^2. The predicted mean vote (PMV) equation has been used in international standards to predict thermal sensation of occupants. It has been widely used by designers to assess indoor thermal environments [8] .

The PMV equation uses the four environmental parameters; air temperature (˚C), mean radiant temperature (˚C), air velocity (m/s), relative humidity (%) and two personal variables (clothing insulation (Clo) and metabolic rate (Met)). The aforementioned when calculated, forms an index that can be used to predict the thermal comfort levels within a space. The index provides a score that corresponds to the ASHRAE thermal sensation scale and represents the average thermal sensation felt by a group of people in a space [9] [10] . Table 1 shows the seven point ASHRAE thermal sensation scale.

Overall, thermal comfort field studies [8] [11] have suggested that the PMV model does not always accurately predict the actual thermal sensation of occupants. Researchers have cited measurement error and contextual assumptions as the two contributing factors to the discrepancies [12] . Measuring of the physical variables with reliable instrument, clothing insulation, activity levels, individual differences, building differences, outdoor climate, behavioural and psychological adaptation are all factors that have been mentioned by thermal comfort researchers as factors contributing to the bias nature of the PMV model [12] -[14] . The PMV model prediction has been found by a majority of researchers as a better predictor in mechanically ventilated buildings [15] and [16] . In the case of naturally ventilated buildings, differences have been observed between the thermal sensations reported by the occupants and that of the PMV model [17] . Studies have shown that the PMV either underestimates or over-estimates the thermal sensation of occupants. Reference [18] and [19] for instance observed the underestimation of thermal sensation by PMV model in their respective studies.

Table 1. ASHRAE thermal sensation scale.

−3	−2	−1	0	+1	+2	+3
Cold	Cool	Slightly Cool	Neutral	Slightly Warm	Warm	Hot

Air temperature is often taken as the main design parameter for thermal comfort. Hence it is essential for occupants' well-being, productivity and efficiency [20] . Reference [21] suggested that air temperature alone is a good indicator of thermal comfort. The effects of other environmental parameters on thermal comfort have also been studied by [22] as cited in [5] .

Reference [23] in his study asserted that, heat gain through the exterior window accounts for 25% - 28% of the total heat gain. Moreover, the thermal capacities of various materials respond differently on incident solar radiation [24] .

Orientation and spatial organization also affect the ability of a building to ventilate and receive solar radiation [25] . To minimize solar heat gain and maximize ventilation, buildings should be orientated with their longer sides intercepting prevailing winds and the shorter sides facing the direction of the strongest solar radiation. The positive effects of shading on thermal comfort have been stated by [26] and [27] .

Reference [28] studied thermal comfort and workplace occupant satisfaction in low energy office buildings in Germany. They reported that votes of "just right" and "slightly warm" covered operative temperatures of above 27˚C. Again, they noted that PMV was very wide spread on the sensation scale and only changed slightly, dependent on the class of the subjective votes ("just right", "slightly warm" or "very warm").

Similar studies conducted by [17] in UK summer conditions found a mean temperature value of 23.9˚C in office buildings with a range of 21.6˚C to 26˚C. Additionally, there was an average PMV value of −0.25 with a range of −1.6 to 0.5. Reference [16] also found lower mean temperatures of 23.8˚C in offices he studied in London.

Very few studies have been conducted in the area of thermal comfort in Ghana [29] [30] . There is therefore a need to build a growing literature of the subject. Thermal comfort and indoor thermal environment can have an important impact on productivity [31] . Accordingly the aim of this study is to evaluate thermal comfort levels in naturally and mechanically ventilated offices using the PMV-PPD. Eventually, all the stakeholders in the building industry will be informed on best practices that can improve upon thermal comfort in office buildings.

3. RESEARCH METHODOLOGY

Thermal comfort of the offices was studied by means of quantitative survey along with questionnaire for building occupants. The thermal comfort index (PMV) which formed the basis of the study is considered to be applicable within an environment for the following reasons; Fanger's investigation was based upon a wide range of environmental conditions, including conditions when air temperature >30°C, and relative humidity up to 70%. In addition, a moderate thermal environment is deemed to have been achieved when PMV values range from −3 as cold to +3 as hot. Figure 1 below illustrates the functional relationship between PMV and PPD. At PMV neutral (0), 5% of the occupants are still dissatisfied.

Four buildings (Table 2) situated within the central business district of Accra were selected for the studies. The selected buildings are representative of current design trends in Ghanaian high-tech office construction due to their exposed glazing.

3.1. Participants in the Study

In all, 195 occupants in all 4 buildings participated in the study. All participants were workers on the floors monitored and had stayed in their current offices from between a year to ten years. The respondents in total consisted of 80 females and 115 males. Greater majority of the respondents (131) were between the ages of 25 - 35 years, while 8 were under 25 years, 29 participants were between 36 - 45 years and 37 were above 45 years.

A field study was carried out from the 14th April, 2012 to 31st January, 2013, representing both the raining and the dry seasons in Ghana. Subjective measurements (in form of a thermal comfort questionnaire) and objective data were recorded during the study period. The objective data made use of loggers (onset hobo sensors for environmental parameters) which were installed at various points in the offices. Indoor temperature and relative humidity levels were measured at 10 minutes intervals for the study period. Altogether, a total of 714,286 data set points were measured and used for the study.

Table 2. General overview of selected buildings.

Building Code	Fl. Area (m^2) (No. of floors)	Location	Orientation	Floors monitored	Thermal controls	Shades	Windows	Use
P.T.	10263 (13)	Accra, CBD	East-West	5th and 6th	Air-conditioned	Internal, Manually controlled	Fixed curtain wall	Multi-purpose
W.T.C.	14556.78 (15)	Accra. Ridge	South-East	14th and 15th	Naturally ventilated	Internal, Manually controlled	Operable	Multi-purpose
H.T.	9340.86 (15)	Accra, Ridge	East-West	10th, 11th and 12th	Air-conditioned	Internal, Manually controlled	Fixed curtain wall	Multi-purpose
R.T.	14355.68 (15)	Accra, Ridge	South-East	7th and 8th	Air-conditioned	Internal, Manually controlled	Limited Operability	Multi-purpose

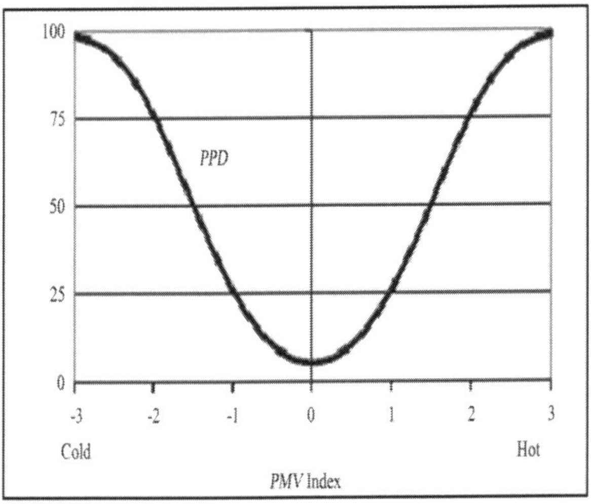

Figure 1. The PMV-PPD curve. **3.2. Experimental Design**

3.1.1. Subjective Measurements

Participants were asked to rate their thermal sensation on the ASHRAE seven-point scale. This was to gather information on how the occupants feel within their office spaces, thus their Actual Mean Votes (AMV). These were then compared to the Predicted Mean Votes (PMV). As part of the questionnaire, occupants were also asked to put down what they normally wore to work. This was used to find the corresponding clothing value from a clothing checklist [17]. Thermal insulation of the chairs of participants (0.1) was added to the clothing insulation for each respondent. Metabolic rate of 1.2 mets which is proposed in ISO 7730 (2005) and asserted in [31] was used as occupants in the offices engaged in sedentary activities.

3.1.2. Objective Data

Air temperature (T) and relative humidity (R.H) values were monitored with onset hobo sensors, whilst the hot wire anemometer was used to measure the air velocity (A.V). Measurements were taken at a height of 1.1 m above the floor, which represent the height of an occupant when seated. The accuracy of the measuring instruments is shown below (Table 3).

Table 3. Measuring equipment technical data.

Parameter	Measuring Range	Accuracy
Air temperature (T)	−20°C to 70°C	±0.4°C
Relative humidity (R.H)	5% to 95%	±3%
Air velocity (A.V)	0.1 to 25.0 m/s	±5% ±0.1 m/s

3.2. PMV Calculation

The software, PMV calc v2 was used to calculate the Predicted Mean Vote (PMV) and the Predicted Percentage of Dissatisfied occupants (PPD) after [32] . This was achieved by inputting the clothing values, metabolic rates of the occupants, air velocities, indoor temperatures and relative humidity values of the occupied spaces into the software. The output results were tabulated and analyzed.

3.3 Statistical Analysis of Data

Greenline software was used to launch and download the files from the data loggers. The downloaded values were screened and exported to an MS Excel file. In Excel, the text files were screened and built together in monthly tables (comprising of all the days within the month). Formulae sheets (excel working sheets) were generated to produce mean hourly values. Output values were generated considering only the working hours from 8:00 am to 5:00 pm. The generated values were analyzed using descriptive statistics.

4. RESULTS AND DISCUSSION

The results for the study are presented. They provide clues on the thermal comfort levels in the studied offices.

4.1. Outdoor Conditions

The lowest mean outdoor temperature measured within the study period was 26.2˚C (in August) corresponding to 76.2% relative humidity (R.H) with the highest being 29.8˚C recorded in November and 74.4% R.H. The highest R.H was however recorded in July with a value of 78.9% with January recording the lowest of 72.6%. With the values recorded, the two main climatic seasons in Ghana is clearly indicated.

4.2. Indoor Measurement Results and Survey Responses

Table 4 shows mean, minimum and maximum values of all indoor environmental parameters. PMV of occupants with its corresponding PPD is illustrated. The Actual Mean Votes (AMV) of the respondents was also determined. Based on the findings, the average Relative humidity values recorded within the MVB's were 54%, 59% and 51.2% as against 71.6% in the NVB. The subjective analysis of humidity is also presented in Figure 2. The respondents were satisfied with the humidity levels in the offices though it is known that humidity has little or no effect on thermal comfort when within 6% - 90% [33] .

From Table 4, it can be deduced that the mean air temperatures for the MVBs' falls within the ASHRAE's acceptable temperature range of 23˚C - 26˚C [2] . Albeit, many studies have concluded that the comfort temperature is higher in tropical regions, since humans have the ability to acclimatise [34] . Reference [35] reported that the comfort band for Malaysia for all building types is between 23.6˚C - 28.6˚C. Air velocity from Table 4 ranges from 0.2 - 0.5 m/s with an average of 0.2 m/s, a result which correlates with [36] studies. In their studies, average air velocity was found to be 0.16 m/s with most samples having air velocity of lower than 0.2 m/s [36] .

In the World Trade Centre (W.T.C) which is a naturally ventilated building (NVB), average PMV recorded was 1.5 (warm) which corresponded to 51.4% PPD. Meanwhile, the indoor temperature ranged from 26.7 - 29.5˚C. These values are in sharp contrast with [17] studies. In their study, indoor temperature values ranged from 21.6˚C - 26˚C with an average of 23.9˚C.

PMV value of −0.25 was also found: making W.T.C warmer with a PMV of 1.5.

The finding in the NVB also contrasts [6] study in the Ghadames, Libya. In their work, the same metabolism rate and clothing values used in this study was used to calculate the PMV values for occupants in naturally ventilated buildings. The PMV was found to be 2.7 (hot). The difference in the two values could be attributed to the climate characteristics of Libya. The climate (Ghadames) is characterized by high air temperature, high solar radiation, low rainfall, low humidity and many sandstorms.

Shape and orientation to breezes, vertical shafts to promote air flow, wing walls and overhangs in the direction of winds have to be considered for naturally ventilated buildings [30] . In addition, intelligent systems and system controllers have to be installed in buildings since they have positive effect on thermal comfort [37] . The installation of fans (with a temperature reduction capability of 2˚C [30] could be installed within the offices to reduce the high temperature values as short term improvement mechanism in the free-running building.

The H.T building also recorded an average indoor temperature of 26.1˚C with a range of 24.8˚C - 27.7˚C. The aforesaid could be attributed to the orientation of the building (east-west) with its unprotected glazing: confirming the assertion of a number of studies that heat gain through the exterior window accounts for 25% - 28% of the total heat gain within a space [23] and [38] . A PMV of 0.8 was recorded whiles an AMV of −0.5 was indicated by the respondents. Again though H.T is a MVB, the slightly warm PMV indication is as a result of it orientation and the lack of shading. What is worse, the glazing has also been fixed on the east-west sides of the building.

Respondents in the H.T building during the afternoons take long breaks (1 - 1.5 hrs) outside their offices (35%) whiles those with single occupancy set their air-conditions at 18˚C (19%) instead of the general 20˚C - 24˚C by management.

Other respondents (37%) also felt comfortable within their clothes and yet again, 9% are indifferent.

Table 4. Statistical summary of indoor measurement results.

Parameters	Unit	Premier Towers	World Trade Centre	Heritage Towers	Ridge Towers
Air temperature-Mean value (min-max)	°C	24.5 (22.5 to 25.9)	28.4 (26.7 to 29.5)	26.1 (24.8 to 27.7)	24.3 (23.9 to 25.1)
Relative humidity-Mean value (min-max)	%	54 (48 to 58.2)	71.6 (70 to 73.3)	59 (54 to 68)	51.2 (50 to 52.6)
Air velocity-Mean value (min-max)	m/s	0.1 (0 to 0.2)	0.35 (0.2 to 0.5)	0.11 (0 to 0.21)	0.1 (0 to 0.2)
Clothing insulation	clo	Cotton working shirt with long sleeves/short sleeve, working trousers/skirt, 3piece suit. Average clothing insulation value used (0.8 clo).			
PMV Mean value (min-max)	-	0.3 (−0.02 to 0.6)	1.5 (1.0 to 1.8)	0.8 (0.4 to 1. 2)	0.2 (0.1 to 0.4)
PPD Mean value (min-max)	%	11 (5.5 to 18.3)	51.4 (27.6 to 67.3)	20.24 (9 to 38.1)	6.28 (5.2 to 9)
AMV Mean value (min-max)	-	−0.5 (−3 to 2)	0.5 (−2 to 3)	−0.5 (−3 to 2)	−1.5 (−3 to 0)

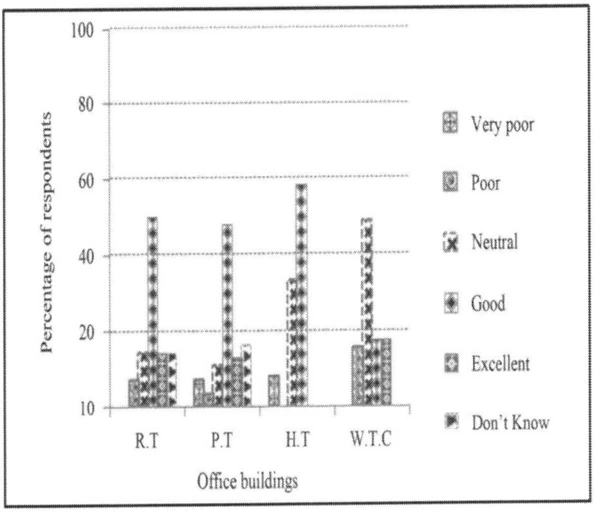

Figure 2. General feeling concerning Humidity.

4.3. PMV and AMV Comparison

In Table 4, PMV calculation for the buildings indicated that the occupants inside the mechanically ventilated buildings were comfortable (averages were all within the central three categories of the thermal sensation scale: −1, 0, 1). The NVB recorded a PMV of 1.5 (between slightly warm and warm). Figure 3 shows the subjective responses (AMV) in all the buildings.

PMV and the average AMV in the NVB (W.T.C) shows wide disparities. Whiles PMV recorded is 1.5, average AMV is 0.5 (Table 4). Significantly, PMV has over-predicted the indoor thermal comfort in the W.T.C by 1 scale unit: A result which is in the range of the findings of [39] . They concluded in their study that there was a discrepancy between PMV and AMV by 1.3 scale units. The over-prediction nature of the PMV has also been recorded by other researchers in their studies [15] [40] . Some reasons for the over-prediction in the current study could be summarised as follows: The fact that the respondent had a couple of adaptive advantages (open windows, use of personal standing fans).

Again, the respondents in the office work in the mornings when the temperature was assumed to be comfortable. During the afternoons where temperature recorded is high, 90% of the respondents are out of the office of various assignments since it's a financial research company.

Another rationale could also be that the respondents had acclimatised to the temperature and humidity values and therefore reported as such.

Additionally, there are a number of variations within the office spaces that could lead to the high value of the PMV which was not the case during the climate chamber experiment. Amongst them include the use of office equipment (laptops, photocopy machines, printers etc.) and the partitions for individual personalised workstations. Some respondents who are located along the windows are very comfortable in the mornings when the windows are opened (53.9%) whiles those within the central part of the office space are uncomfortable (10.2%).

5. CONCLUSIONS

Analysis of data collected and findings from the assessment of thermal comfort in multi storey office buildings have been presented in this paper. The study aimed at evaluating thermal comfort levels of naturally and mechanically ventilated offices in comparison with ASHRAE 55 and ISO 7730 standards.

The study employed Fanger's PMV-PPD model in finding the thermal comfort levels in the office spaces. Following the results and discussion, it is stated that the PMV-PPD model comfortably predicts thermal comfort in MVBs' than in NVBs'. Thus, thermal comfort levels found within the MVB compared well with the ASHRAE and ISO standards. The mean temperature of 25˚C for the MVB falls within ASHRAE's acceptable temperature range of 23˚C - 26˚C (summer comfort). Meanwhile the average subject satisfaction for the MVB of 87.5% is also within ISO 7730's range of 75% and close to ASHRAE's 90% of total subject satisfaction which is not the case for the NVB.

6. RECOMMENDATION

Adaptive principles must be used to enhance the thermal comfort conditions (external shading, reduction in wall-to-window ratio, etc.) in naturally ventilated buildings.

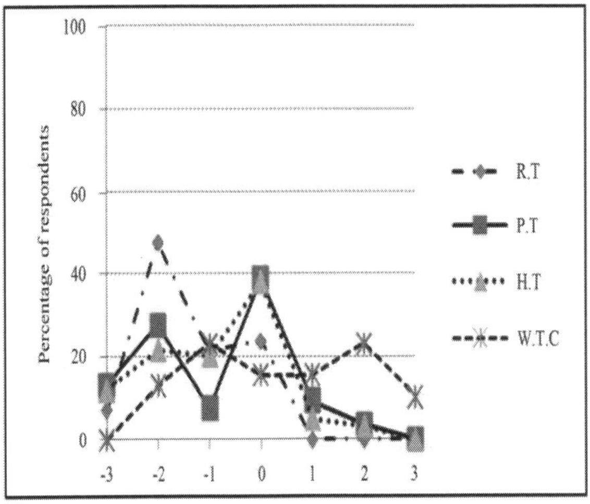

Figure 3. Respondents thermal comfort levels in the buildings.

Further studies should however be conducted to ascertain the reliability of the PMV-PPD to predict the thermal sensation within offices spaces.

Best practices that will lead to the efficient use of air-conditions should be looked into.

REFERENCES

1. Budaiwa, I.M. (2006) An Approach to Investigate and Remedy Thermal-Comfort Problems in Buildings. Building and Environment, 42, 2124-2131. www.Sciencedirect.com

2. American Society of Heating, Refrigerating and Air-Conditioning Engineers (ASHRAE) (2004) Standard 55-2004 & Standard 55-1992,

Thermal Environmental Conditions for Human Occupancy. ASHRAE, Atlanta.

3. ISO 7730 (2005) Moderate Thermal Environments—Determination of the PMV and PPD Indices and Specifications for Thermal Comfort. 2nd Ediyion, International Organization for Standardization, Geneva.

4. Charles, K.E. (2003) Fanger's Thermal Comfort and Draught Models. Institute for Research and Construction. IRC- RR-162, Institute for Research in Construction, Ottawa.

5. Szokolay, S. (2004) Introduction to Architectural Science: The Basis of Sustainable Design. Architectural Press, Oxford, 17.

6. Ealiwa, M.A., Taki, A.H, Howarth, A.T. and Sedan, M.R. (2001). An Investigation into Thermal Comfort in the Summer Season of Ghadames, Libya. Building and Environment, 36, 231-237. www. elsevier.com/locate/buildenv

7. Hyde, R. and Sabarinah, S.A. (2008) Bioclimatic Housing, Innovative Designs for Warm Climates. Earthscan, UK & USA.

8. Humphreys, M.A. and Nicol, J.F. (2002) The Validity of ISO-PMV for Predicting Comfort Votes in Everyday Thermal Environments. Energy and Buildings, 34, 667-684.

9. American Society of Heating, Refrigerating and Air-Conditioning Engineers (ASHRAE) (2001) Thermal Comfort. In ASHRAE Handbook: Fundamentals, ASHRAE, Atlanta, 8.1-8.29.

10. Fanger, P.O. (1970) Thermal Comfort. Danish Technical Press. Copenhagen.

11. de Dear, R., Fountain, M., Popovic, S., Brager, G. and Arens, E. (1993) A Field Study of Occupant Comfort and Office Thermal Environments in a Hot-Humid Climate. Final Report, ASHRAE RP-702, American Society of Heating, Refrigerating and Air-Conditioning Engineers, Atlanta.

12. de Dear, R.J. and Brager, G.S. (2002) Thermal Comfort in Naturally Ventilated Buildings: Revision to ASHRAE Standard 55. Energy and Buildings, 34, 549-561.

13. Morgan, C.A., De Dear, R. and Brager, G. (2002) Climate, Clothing and Adaptation in the Built Environment. In: Levin, H., Ed., Indoor Air 2002:

The 9th International Conference on Indoor Air Quality and Climate, Monterey, 30 June-5 July 2002, 98-103.

14. Nicol, F.G. (2001) Characterising Occupant Behaviour in Buildings: Towards a Stochastic Model of Occupant Use of Windows, Lights, Blinds, Heaters and Fans. Proceedings of 7th International IBPSA Conference, Rio de Janeiro, 13-15 August 2001, 1073-1078.

15. Brager, G.S. and De Dear, R.J. (1998) Thermal Adaptation in the Built Environment: A Literature Review. Energy and Buildings, 27, 83-96.

16. Oseland, N.A. (1995) Predicted and Reported Thermal Sensation in Climate Chambers, Offices and Homes. Energy and Buildings, 23, 105-115.

17. Beizaee, A., Firth, S.K., Vadodaria, K. and Loveday, D. (2012) Assessing the Ability of PMV Model in Predicting Thermal Sensation in Naturally Ventilated Buildings in UK. 7th Windsor Conference: The Changing Context of Comfort in an Unpredictable World, Windsor, 12-15 April 2012, Network for Comfort and Energy Use in Buildings, London.

18. Hong, S.H., Gilbertson, J., Oreszczyn, T., Green, G., Ridley, I and the Warm Front Study Group (2009) A Field Study of Thermal Comfort in Low-Income Dwellings in England Before and After Energy Efficient Refurbishment. Building and Environment, 44, 1228-1236.

19. Han, J., Zhang, G., Zhang, Q., Zhang, J., Liu, J., Tian, L., Zheng, C., Hao, J., Lin, J., Liu, Y. and Moschandreas, D.J. (2007) Field study on Occupants' Thermal Comfort and Residential Thermal Environment in a Hot-Humid Climate of China. Building and Environment, 42, 4043-4050.

20. Adebamowo, M.A. and Akande, O.K. (2010) Indoor Thermal Comfort for Residential Buildings in Hot-Dry Climate of Nigeria. http://nceub.org.uk

21. Heidari, S. and Sharples, S. (2002) A Comparative Analysis of Short-Term and Long-Term Thermal Comfort Surveys in Iran. Energy and Buildings, 34, 607-614

22. Auliciems, A. (1981) Towards a Psycho-Physiological Model of Thermal Perception. International Journal of Biometeorology, 25, 109-122.

23. Al-Najem, A.A. (2010). The Effects of Orientation, Ventilation and Varied WWR on the Thermal Performance of Residential Rooms in the Tropics.http://www.ccsenet.org/journal/index.php/jsd/article

24. Stein, B. and Reynolds, J.S. (2000) Mechanical and Electrical Equipment for Buildings. Ninth Edition, John Wiley and Sons, Inc., New York, 455-475.

25. Koranteng, C. and Abaitey, E.G. (2010) The Effects of Form and Orientation on Energy Performance of Residential Buildings in Ghana, Journal of Science and Technology, 30, 71-81.

26. Chenvidyakarn, T. (2007) Passive Design Techniques for Thermal Comfort in Hot Humid Climate. www.shibaura-it.ac

27. Sumanon, R. (2004) The Unique Angle of Roof Slope Effecting Thermal Comfort in the Traditional Thai House. http://www.arquitecturatropical.org.pdf

28. Wagner, A., Gossauer, E., Moosmann, C., Gropp, T. and Leonhart, R. (2007) Thermal Comfort and Workplace Occupant Satisfaction—Results of Field Studies in German Low Energy Office Buildings. Energy and Buildings, 39, 758- 769. http://dx.doi.org/10.1016/j.enbuild. 2007. 02.013 www.sciencedirect

29. Koranteng, C., Nyame-Tawiah, D. and Quansah, E. (2011) A Psychometric Analysis of Thermal Comfort in Low-Rise Office Buildings in Ghana. Journal of Science and Technology, 31, 76-88.

30. Koranteng, C. and Mahdavi, A. (2010) An Enquiry into the Thermal Performance of Five Office Buildings in Ghana. 10th Revha World Congress on Sustainable Energy Use in Buildings (CLIMA 2010), Antalya, 9-12 May 2010.

31. Daghigh, R., Adam, N.M., Sopian, K. and Sahari, B.B. (2009) Thermal Comfort of an Air-Conditioned Office through Different Windows-Door Opening Arrangements. Building Services Engineering Research and Technology, 30, 49-63.

32. Fanger, P.O. (1967) Calculation of Thermal Comfort: Introduction of a Basic Comfort Equation. ASHRAE Transactions, 73, III4.1-III4.20.

33. Olesen, B.W. and Brager, G.S. (2004) A Better Way to Predict Comfort. ASHRAE Journal, 46, 20-28. http://www.cbe.berkeley.edu/research

34. Xi, T., Li, Q., Mochida, A. and Meng, Q. (2012) Study on the Outdoor Thermal Environment and Thermal Comfort around Campus Clusters in Subtropical Urban Areas. Building and Environment, 52, 162-170.

35. Sabarinah, A. and Steven, S. (2007) The Performance of a Partially Air-Conditioned Apartment Building in Kuala Lumpur. The 24th International Conference on Passive and Low Energy Architecture, Singapore, 22-24 November 2007, 608-614.

36. Huang, L., Zhu, Y., Ouyang, Q. and Cao, B. (2012) A Study on the Effects of Thermal, Luminous, and Acoustic Environments on Indoor Environmental Comfort in Offices. Building and Environment, 49, 304-309.

37. Mohammadi, A. (2007) Modeling Occupants' Control Actions and Their Energy Implications in an Office Building. Institute of Building Physics and Building Ecology. Technology University of Vienna, Austria.

38. Yu, J., Yang, C. and Tian, L. (2008) Low Energy Envelope Design of Residential Building in Hot Summer and Cold Winter Zone in China. Energy Build, 40, 1536 -1546.

39. Doherty, T.J. and Arens, E. (1988) Evaluation of the Physiological Bases of Thermal Comfort Models. ASHRAE Transactions, 94, 1371-1385.

Chapter 7

The Production of Thermal Insulation by Waste Animal and Vegetable Fibres

Binici H and Kekili M*

Kahramanmaras Sütçüimam University, Department of Civil Engineering, Kahramanmaras, Turkey

ABSTRACT

World energy consumption refers to the total energy used by all of human civilization. Since the heating costs in winter, the cooling costs in summer are very high in Turkey, insulation is made to the buildings to reduce these costs. A great part of the raw materials of nsulation materials have been imported. Rapid increase in the population, developments in the technology everyday bring rapid consumption of energy resources with them. Today, the energy consumption has reached serious proportions and precautions have already been started to be taken all over the world. Most of the energy in Turkey is spent in industry and in housings. Energy savings potential is available up to annual $ 7.5 billion by the dissemination of heat insulation applications in Turkey. Energy efficiency should be supported in buildings via heat insulation in order to prompt this potential. In the present study, animal and vegetable fibre doped insulation material has been produced at specified rates. Set out in the findings, research materials waste animal and vegitables fibers shows that it is possible to produce insulation materials enough to mix it with plaster. The thermal insulation coefficient of the obtained samples conforms to the TS EN standards.

KEYWORDS

Animal fibres; Vegetables fibres; Insulation material

1. INTRODUCTION

Since the heating costs the cooling costs are very high in Turkey, insulation is made to the buildings to reduce these costs. A great part of the raw materials of insulation materials have been imported. Rapid increase in the population, developments in the technology everyday bring rapid consumption of energy resources with them. Today, the energy consumption has reached serious proportions and precautions have already been started to be taken all over the world. Most of the energy in Turkey is spent in industry and in housings. While hot water supply lightening and, the use of electrical appliances constitute15% of the energy consumption in buildings, 85% is spent for cooling and heating [1].

Energy savings potential can be made to annual $ 7.5 billion by the dissemination of heat insulation applications in Turkey. Energy efficiency should be supported in buildings via heat insulation in order to prompt this potential. Animal and vegetable fibre doped insulation material has been produced at specified rates. The thermal insulation coefficient of the obtained samples conforms to the standards [2]. In the near future, all new buildings will be built thanks to the passive house and the available buildings will have been renewed in order to meet low energy building standard [3]. In recent years, more emphasis is given to ecological characteristics in the area of building heat preservation. Environmental consciousness is not limited with the energy saving any more, also it only takes place in ecological structure, in other words, it is to be a part of minimum energy input, resource consumption and pollution production, installation and insulation material should be used. It has been determined that the energy resources in the World are consumed away rapidly, and the about 40% of energy consumed in Turkey has been consumed in buildings [4]. 80% of this energy is for heating and more than 65% of the consumed energy in Turkey is imported. Therefore, insulation materials contribute to constructions for heating savings on a considerable amount. Different composites productions have been actualized from vegetable fibres [5]. For this reason, solving the problem of production and the production of composite plates from wheat straws has become successful. It has been determined that chip and fibre boards produced by wheat straw are more advantageous in many ways than those produced from wood. This has increased the demands of using agriculture wastes by the composite panel producers in Canada who makes researches especially about this issue. In Turkey that has great wheat straw potential, making studies for this kind of researches and applications is inevitable. On the other hand, the production of sunflower has been carried out at significant amounts. Sunflower stalks emerging after the production is very serious problem for our farmers. There are 2.500.000 tons sunflower stalks per year as waste [6]. On the other hand, the feature of light and new insulation material produced by cotton wastes and textile ash has been investigated. The experiments of compressive strength and thermal conductivity of the block insulation material have been carried out in accordance with ASTM and Turkish Standards. As a result of experiments; it has been emphasized that the light insulation materials produced have good insulation features and are light composite used in the production of brick, wall and ceiling panels. The coefficients of thermal conductivity of the composites obtained by using natural fibres have been determined [7,8]. It has been determined that the coefficients of thermal conductivity of the composites mentioned are lower than those that are produced by artificial fibres, and that they are economic, don't harm the environment and also have very good mechanical features.

The use of natural fibres in insulation is related to the ecological building sector [9], where selection of materials is based on factors including recyclable, renewable raw materials and low resource production techniques. In addition to cellulosic insulations have a higher moisture regainthanin organic materials and therefore only cellulosic materials are recommended for old timbered houses [10,11]. One reason for the relatively low utilisation is higher price of fibrein sulations compared to mineral wool. On the other hand, these insulation materials has low energy demand in production, potential forrecycling and positive effect on indoorair [12]. The thermal performance of bio-based materials has been investigated and confirmed generally as competitive with conventional insulations in some studies [13,14]. Considerable data is available concerning the technical properties of conventiona linsulations [15,16]. The implementation of agro-industrial residues in the development of composite materials for building trades is now a research of growing interest [17,18]. One of the major importance of this industry for the next decade is to improve the energy performance of existing building staking into account the increasing social emphasis on issues of the environment, waste disposal, and the depletion of non-renewable resources. The motivation includes cost, mechanical, thermal and acoustical performance, weight reduction, and environment concerns [19]. Thermal insulation is known to play a vital role in saving energy by reducing the rate of heat transfer [20]. Among them, those obtained from agriculturalby-products have numerous advantages as they are eco-friendly and renewable. Furthermore, the high fiber content of some of them authorizes the reinforcement of composites [21]. The bio-sourced material obtained had higher thermal conductivity and mechanical strength compared to cement alone [22].

As it is understood from the previous studies, it is of great importance to assess the wastes in the world any more. In this regard, the purpose of this study is to produce appropriate insulation materials that will assess the wastes .This study aims to produce a fireresistant insulation material which has much lower heat transmission coefficient. Thus, both the local people will benefit from the wastes and employment will be provided by mass production and also they will contribute to the economy. On the other hand, since the studies made by thermal insulation are limited, the studies made by us will constitute to literature

2. MATERIALS AND METHODS

2.1 Material

Vegetable and animal fibres: Sunflower: Sunflower is one of the most important oil plants in the world and in Turkey and it is mostly produced for oil in our country. As it is understood from the previous studies, it is of great importance to assess the wastes in the world any more [6]. In this regard, sunflower stalk will be used in the production of insulation material in this study as vegetable fibre which is found plentifully in Elbistan Plain. In order to offer solution directly to components and elements produced from the plaster, the raw material of which is found commonly in our country, shortening the construction period, mitigating construction, preventing place and material losses, planning and bring flexibility of use, reducing maintenance and finally, the purpose of reducing costs of construction and usage, our country should find more application area in terms of construction sector [23]. Nine million tons sunflower stalks have been determined as waste in Turkey as of 2012. Approximately 5 thousand insulation materials can be produced by these wastes.

Wheat stalk: The wheat stalks (stubble) that are found abundantly on the region and burnt by farmers after harvest have been collected appropriately. This has aimed both to make it win for economy and to prevent it from pollution. Sunflower and wheat stalks and agriculture wastes in our country are pretty much and heat conservatism and void ratios are the most important parameters of providing insulation in the layers of composite. Evaluating these wastes will both contribute to the economy of the country and will provide to get ahead of air pollution as a result of burning these wastes.

Animal fibres (goat hair and sheep wool): Every year about 35 million tons natural fibres have been produced including animal and vegetable origin in the world. While vegetables origin natural fibres produced at industrial level are essentially cotton, coconut fibre, linen, hemp, jute, the animal origin fibres are wool, silk, mohair and Kashmir. They are Angora rabbit hair, goat upper coarse fibres and Asia and the fibres obtained from the South America camels. By using these fibres on different production areas such as textile, automotive, furniture, agriculture and handcrafts, they contribute to the country economy at significant level in the countries where they are produced and/or they are demanded. However, during the last half-century, petrochemical fibres such as acrylic, nylon, polyester and polypropylene taking part instead of natural fibres in different production areas have caused the production and usage of vegetable and animal fibres to be regressed at significant level. Today, fibre has been obtained for trade purposes from approximately 9 different animal kinds in the world. Those except silk worm fall into the group of mammals. In Turkey, mainly wool, mohair, silk, goat upper coarse hair, and very low levels Kashmir and Angora rabbit wool production have been produced [24,25]. Although some of the animal fibres mentioned above have continued in the usage of textile area, the fibres like goat hair are in the situation of waste. In this study, goat hair and sheep wool have been used as fibre.

The duty of goat hair in the composite which is collected from farms is to increase the flexural strength of samples and to provide tensile stress on the surfaces. These materials that have effect to the thermal insulation are not used today or used very little. While sheep wool continues to be valued in some sectors even if just a bit, goat hair is in the situation of waste.

2.2 The materials providing binding and fire-resistant

Vermiculite: Vermiculite long known as a kind of trioctahedral mineral mica is divided into sheets with rapid heating and takes shapes resembling a small worm. The term of vermiculite has been derived from Latin vermıculare by using this feature and has been used for minerals resembling to mica and that its layer load is bigger than simectic and that is expanded 2:1. In the Nature: according to the environment and to the formation, four type have been met including land, autogenic, metamorphic and macroscopic [26]. For the first time, the presence of water molecules has been determined among the mica layers by working the crystal structure of vermiculite. The most important vermiculite deposits discovered in Turkey are Sivas-Yıldızeli and Malatya regions. The reserve calculations indicate 6-7 million tonnes of probable reserves in these occurrences. Today, although the production of vermiculite has increased at a significant amount, it has not reached to the production of developed countries because of not knowing the usage areas. The heat resistance of vermiculite will be determined by this study, and will contribute to production and usage areas.

2.3 Plaster

Gypsum is a mineral, the chemical composition of which is calcium sulphate. Gypsum ($CaSO4+2H2O$) is called to the type having two molecule crystal water in its combination. The main purpose of using plaster is to be binding as well as to provide high temperature resistance. Since it is cheap, plaster has been proffered in the work.

3. METHOD

Samples production: The composites in the work consist of fibres, wastes, main equipment and binding parts .Although the industrial usage of the vermiculite in the natural situation is limited, processed vermiculite has widespread usage area for especially hating and insulation. Although the fibres and wastes found in the Project are not fire-resistance, fire resistance has been provided in the vermiculated mixtures used in surface. The samples produced in preliminary works have been left in fire for 4 hour and deformation has been observed against burning. The samples have preserved their layers and have not been flamed. In the later stages of the work, the waiting times in the fire have been increased and its maximum durability has been observed. A material having much higher fire-resistance in comparison with other insulation materials in the market has been produced. Primarily, the animal fibre used has been brought to the appropriate size by being cut. Then, vermiculite has been added in accordance with the rates given in Table 1. Plaster as binder has been mixed in mixer homogeneously with water (Figure 1). After completing the mixture, moulding starts and the first layer has been poured by compressing it finely and then the exterior of sunflower stalk prepared beforehand has been sprinkled as intermediate layer, the mixture has been compressed by placing it as upper layer. After keeping the sample one day, it has been revealed and has been subjected to heat-treatment and the water and humidity in it have been purified. The composites have been designed both as laminated and granulated composite. The samples produced have been given in Figure 2.

3.1 Applied tests

Specific bulk density and water absorption rate: Unit weight of the insulation materials have been made in accordance with TS EN 2823 and water absorption values have been in accordance with ASTM C 67-03[27].

Ultrasonic sound penetration: There is a significant relationship between the speed of ultrasonic waves and the density of materials [28]. As the amount of space in the material increases, the speed of supersonic wave becomes less. After calculating in how much time the supersonic wave transmitted from a surface of the material block to inside have been passed, the speed of the wave is calculated as follows:

$$V=(S/t)\times 10^6$$

here; V=P speed of wave (kilometre/second) S is the distance (kilometre) between the surface that material block has sent supersonic wave and the surface that the wave has been received, t=P is the time from the material surface that the wave has sent till the surface received (microsecond).The ultrasonic sound transmission experiment have been made on all the samples and the sound transmission speed has been found.

Fire test: After drying in a drying oven selected from six samples, 75, 100, 125 and 150°C temperature have been exposed for 10 minutes. The weight changes of these samples in these temperatures have been determined.

Thermal insulation coefficient: The composite insulation material produced has been measured by KEM brand QTM-500 brand thermal measuring device and its thermal insulation coefficient has been measured. The experiment of determination of thermal conductivity has been made in accordance with ASTM C 1113-90 [29].

4. RESULTS AND DISCUSSION

4.1 Unit weight and water absorption values

The unit weight values of the composites have been given in Figure 3 and water absorption values have been given in Figure 4. It has been found that unit weights of the organic fibre added samples are lower than those that have pure fibre. On the other hand, it has been founded that the unit weight of the fibrous wheat stalk are slightly lower than those that have sunflower doped fibre. As the fibre contribution rate has increased, unit weight values have decreased. The water absorption rates of previous ones have showed differences. As the fibre contribution rate has increased, the water absorption rates have increased. The reason of this is the spaces consisting between fibres and main phase. However, the water absorption rates conform to the standards.

4.2 Ultrasonic sound speed

The Ultrasonic sound of the composites has been given in Figure 5. The Ultrasonic sound speeds of organic doped fibre samples have been found less. The Ultrasonic sound transmission speeds of animal doped fibre samples have been found similar. Naturally, sound absorption capacity of porous materials will be more. These materials have the capacity to provide sound insulation.

4.3 The Thermal conductivity coefficient

Thermal conductivity coefficients of the samples produced in the scope of work have been given in Figure 6. The thermal conductivity coefficients of the samples that aren't organic doped fibre have been found bigger than 0.1. These samples can be accepted as insulation material according to standards. On the other hand, the thermal conductivity coefficients of organic doped fibre have been found less. As the fibre doped rate has increased, thermal conductivity coefficient has declined.

4.4 Fire resistance

Weight losses obtained in the experiments made in order to determine fire resistance by the samples have been given in Figure 7. The reason of vermiculite coefficient is to increase the fire resistance. As expected, this period has extended a little more. The burning period of organic doped fibre samples has been gotten longer from the animal doped fibre samples. However, harmful fumes from the materials have not been observed during fire in both groups. As the oven temperature has increased, naturally, weight loss has increased in all samples. A significant weight loss has not been observed in 75°C temperature.

5. CONCLUSION

It has been seen that the layered samples have less than grained samples in the produced samples in the thermal conductivity coefficients. In the samples that vegetables fibres are used, it has been determined that tensile stresses have been met and they have prevented the dispersal of the samples completely. Since the thermal conductivity coefficients of most of the samples are below 1.00, the composites produced can be used as insulation materials. These materials will be evaluated and a local insulation material has been produced. These new insulation materials can be declined about insulation material. The sunflower and wheat stalks in the waste case will provide important contributions to economy. The cost price of insulation materials to be produced by these wastes can be more economic according to the closest pers.

Table 1: Mix proportion of samples (g).

Sample	Vermikulite	Sheep wool	Goat hair	Stubble	Sunflower Stalk	Plaster	Water
SI	300	20	-	-	-	100	300
S2	300	-	20	-	-	100	300
S3	300	20	-	-	20	100	300
S4	300	-	20	-	20	100	300
S5	300	20	-	-	30	100	300
S6	300	-	20	-	30	100	300
S7	300	20	-	20	-	100	300
S8	300	-	20	20	-	80	300
S9	300	20	-	30	-	80	300
S10	300	-	20	30	-	80	300

Figure 1. Samples preparation process.

Figure 2. Produced samples.

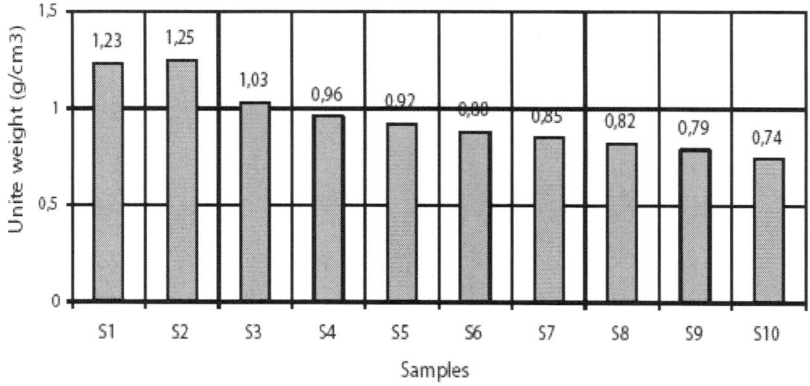

Figure 3. Unite weight of samples.

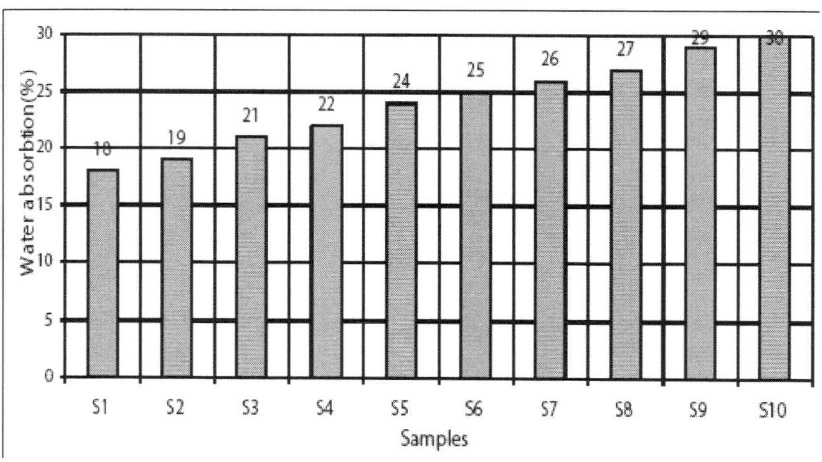

Figure 4. Water absorption of samples

From the fire test, it has been seen that only animal fibres on the surface has been burnt but its surface deformation has not occurred. During experiment, any flame burning has not been observed. Fume cannot be produced in this period; there will no fume-poisoning deaths that are the main reason of the deaths in the fires. Since it is taken into consideration that the materials mentioned in the work is fire-resistance, it can be said that it is technologically superior to others.

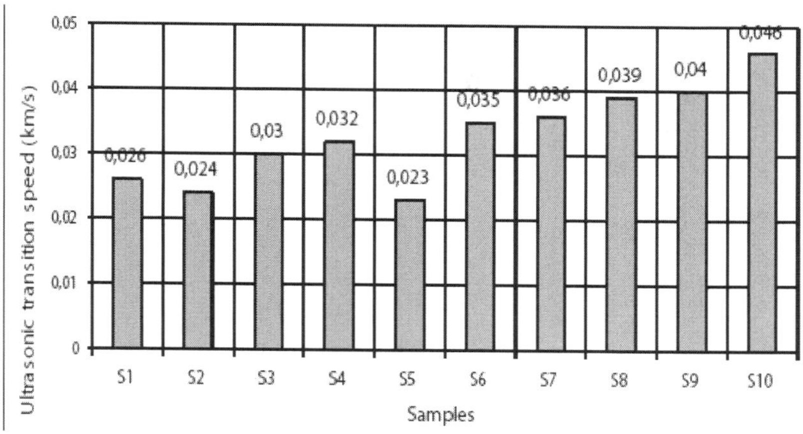

Figure 5. Ultrasonic transition speed of samples.

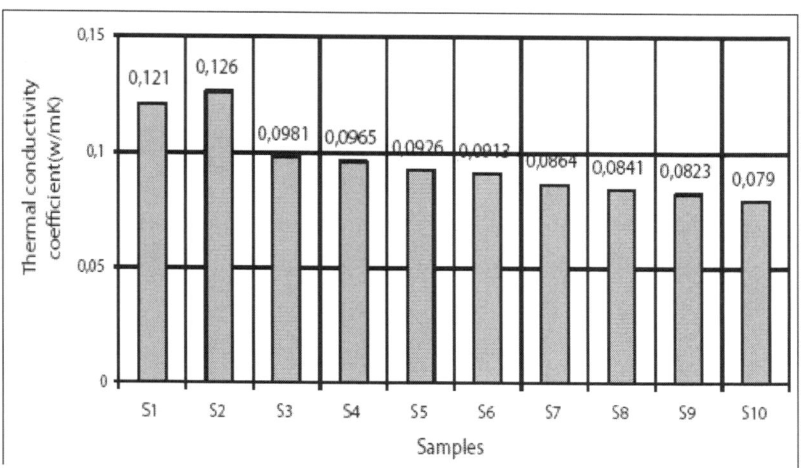

Figure 6. Thermal conductivity coefficient of samples.

Finally, the most important result of this study is that waste materials satisfactory insulation material is produced entirely of animal and organic fibers.

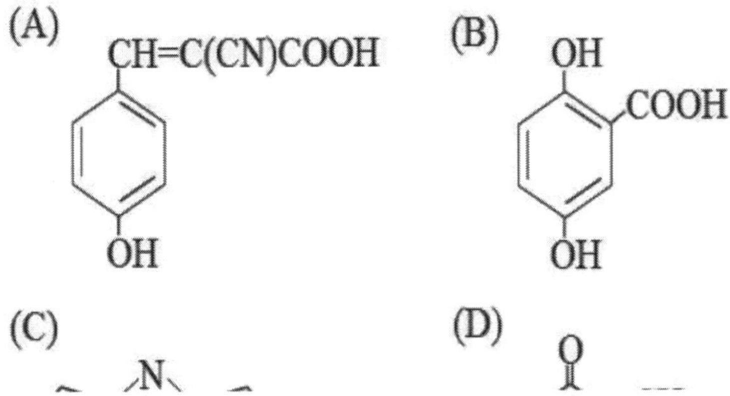

Figure 7. Mass loss of samples after fire test.

REFERENCES

1. de Wilde P, Voorden MV (2003) Providingcomputationalsupportfor the selection of energysavingbuildingcomponents. Energy and Building 36:749-758.

2. Aksoy TU (2008) Investigation of the effect on overallheattransitioncoefficient and heatloss of wallapplications with sandwich and aeratedconcrete. Erciyes UniverstyJournal ofScience 24:277-290.

3. BergeB (2009)The Ecology of Building Materials (2ndedn.)CRC Press,United States.

4. Thorsnes P, Bishop T (2013) The value of basicbuildingcodeinsulation. EnergyEconomy 37:68-81.

5. Mengeloglu F, Alma MH (2002) Wheat stems using composite panel production. Kahramanmaras Sutcu Imam Univrsty Journal of Engineering Science 5:37-48.

6. Binici H,Eken M, Dolaz M, Aksoğan O, Kara M (2014) An environmental friendlythermal insulation material from sun flowerstalk, textile waste and stubble fibres. Construction and BuildingMaterials 51: 24-33.

7. Binici H, Aksogan O (2015) Engineering properties of insulation material made withCotton waste and flyash. Journal of Material Cycles and Waste Management 17:157-162.

8. Monika M, Ramaniah K, Ratnaprasad AV, Mohana RK, Hema CRK (2012) Thermal conductivity characterization of bamboo fibre reinforced polyester composite. J Mater Environ Sci 3:1109-1116.

9. Murphy DPL, Behring H, Wieland H (1997) The use of flax and hemp materials for insulating.Proceedings of flax and other bast plants symposium 30: 79-84.

10. Heier L, Heintges M(1999) Produktion von Industriefaserlein. Landtechnik 54:174-175.

11. Institut N, Karus M (2003) Daskleine Hanf-Lexikon. Verlag DieWerkstatt, Gottingen, Germany.

12. Ringleb A, Schulz H (1996) Der Dammstoffaus Lein—eineernstzunehmende Alternative,Landtechnik 51:40-41.

13. Turner WC, Malloy JF (2001) Thermal insulation handbook. Krieger Publishing Company, FL, USA.

14. Strother EF, Turner WC (1990) Thermal insulation building guide. Krieger Publishing Company, FL, USA.

15. Cole RJ (1999) Energy and gren house gas emissions associated with the construction of alternative structural systems. Build Environ 34: 335-348.

16. Gustavsson L,Sathre R (2006) Variability in energy and carbon dioxide balances of wood and concrete building materials. Build Environ 41: 940-951.

17. Morel JC,Mesbah A, Oggero M, Walker P (2001) Building houses with local materials: meansto drastically reduce the environmental impact of construction. Build Environ 36:1119-1126.

18. Mati-Baouchea N, Baynast H,Lebert A, Suna S, Lopez-Mingo CJS, et al. (2014) Mechanical, thermal and acoustical characterizations of an insulating bio-based composite made from sunflower stalks particles and chitosan. Industrial Crops and Products 58: 244-250.

19. Glé P,Gourdon E, Arnaud L (2012) Modelling of the acoustical properties of hempparticles. Constr Build Mater 37: 801-811.

20. KeynakliO (2012) A review of the economical and optimum thermal insulation thickness for building applications. Renew Sust Energ Rev 16: 415-425

21. Liu Z, Erham SZ, Akin DE, Barton FE (2006) Green composites from renewable ressources: preparation of epoxi dizedsoybeanoil and flax fiber composites. J Agric FoodChem 54: 2134-2137.

22. Korjenic A, Petranek V, Jiri Z, Jitka H (2011) Development and performance evaluation of naturalthermal insulation materials composed of renewable resources. EnergyBuild 43:2518-2523.

23. PatnaikaA,Mlando M, Sudhakar M, Anton B, Rajesh DA (2015) Thermal and sound insulation materials from waste wool and recycled polyester fibers and their bio degradationstudies. Energy and Buildings 92: 161-169.

24. Briga-Sa A, Da Teixeira N, Pinto J, Caldeira F, Varum H, et al. (2013) Textile waste as an alternative thermal insulation building material solution. Construction and Building Materials 38:155-160.

25. Bolat M (2005) Use of Biomass Sources for Energy in Turkey and a ViewtoBiomassPotential.Biomass and Bioenergy 29: 32-41.

26. Guler C,Bektas I, Kalaycıoglu H (2006) The experimental particle board manufacture from sunflower stalks. Forest Production Journal 56:56-60.

27. ASTM C 67-03 (2003)Standard test methods for sampling and testingbrick and structuralclaytile. American Society for Testing and Materials, Philadelphia, PA.

28. ASTM C 597(1994) Standart Test Method For Pulse Velocity Through Concrete. Annual Book of ASTM Standard.

29. ASTM C 1113-90 (1990) Test method fort hermal conductivity of refractoriesby hot wire. Annual Book of ASTM Standard.

Chapter 8

High Energy Performance with Transparent (Translucent) Envelopes

Luis Alonso, César Bedoya, Benito Lauret and Fernando Alonso

E.T.S.A.M. School of Architecture and School of Computing (UPM) Spain

1. INTRODUCTION

Energy efficiency is coming to the forefront in the architecture, as, apart from the significance of a reduced environmental impact and increased comfort for users, the current energy crisis and economic recession has bumped up the importance of the financial cost of energy.

Since the Kyoto Protocol was signed in 1997, governments all over the world have been trying to reduce part of the CO2 emissions by tackling building "energy inefficiency". In Europe today, the tertiary and housing sectors account for 40.7% of the energy demand, and from 52 to 57% of this energy is spent on interior heating (Willems & Schild, 2008). The new world energy regulations, set out at the European level by the Commission of the European Communities in the First Assessment of National Energy Efficiency Action Plans as required by Directive 2006/32/EC on Energy End-Use Efficiency and Energy Services, (Commission of the European Communities, 2008) indirectly promote an increase in the thickness of outer walls, which, for centuries, have been the only way of properly insulating a building.

The use of vacuum insulation panel (VIP) systems in building aims to minimize the thickness of the building's outer skin while optimizing energy performance. The three types of vacuum chamber insulation systems (VIS) most commonly used in the construction industry today –metallized polymer multilayer film (MLF) or aluminium laminated film, double glazing and stainless steel sheet or plate (Willems & Schild, 2006) –, have weaknesses, such as the fragility of the outside protective skin, condensation inside the chamber, thermal bridges at the panel joints, and high cost, all of which have a bearing on on-site construction (Baetens et al., 2010).

Apart from overcoming these weaknesses and being a transparent system, the new F^2TE^3 (free-form, transparent, energy efficient envelope) system that we propose has two added values. The first is the possibility of generating a structural skin or self-supporting façade. The second is the possibility of designing free-form architectural skins. These are research lines that the Pritzker Architecture Prize winners Zaha Hadid, Frank Gehry, Rem Koolhaas, Herzog & de Meuron, among many other renowned architects, are now exploring and implementing.

To determine the feasibility of the new envelope system that we propose, we compiled, studied and ran laboratory tests on the materials and information provided by commercial brands. We compared this information to other independent research and scientific trials on VIPs, such as Annex39 (Simmler et al., 2005), and on improved core materials, such as hybrid aerogels and organically modified silica aerogels (Martín et al., 2008), conducted by independent laboratories like Zae Bayern in Germany (Heinemann et al., 2009), the Lawrence Berkley Laboratory at the University of California (Rubin and Lampert, 1982) or the Technical University of Denmark (Jensen et al., 2005).

After studying the results, we discovered valuable innovative ideas that we exploited to design the new high energy efficient envelope that should outperform the elements now on the market.

The remainder of the chapter is structured as follows. In Section 2 (Research) we explain the rationale of the epistemological study of the system, and how the experimental study combining computer simulations and empirical trials was run. In Section 3 (Experimental study) we describe the design of the proposed envelope system (F^2TE^3), explaining the solutions adopted in this new system and the improvements on other existing systems. Section 4 (Free-Form, High Energy Performance, Transparent Envelope System (F^2TE^3) Design) analyses how the F^2TE^3 system overcomes the weaknesses detected in existing VIP systems. Finally, Section 5 (Conclusion) discusses final conclusions.

2. RESEARCH

Today's architectural vanguard demands a building system such as is proposed in this research: a lightweight, variable geometry, seamless high energy performance system that also permits the passage of natural light and backlighting.

No system combing all these features exists as yet, and similar systems are not absolutely free form and translucent, are not seamless and/or have a very limited thermal response.

From the viewpoint of energy performance, of the three types of translucent insulations that there are on the market (plastic fibers, gas and aerogel), we found that the insulation that best meets the needs of the new system that we propose is aerogel (Ismail, 2008; Wong et al., 2007).

Aerogel and nanogel (granular silica gel) have four advantages for use as thermal insulation in translucent panels:

a. Transparency: aerogel is comparable to glass in terms of transparency (Moner-Girona et al., 2002). Monolithic aerogel light transparency can be as high as 87.6%, even greater than what some gas-insulated glasses can offer. Granular nanogel offers a greater translucency than any other traditional insulation material (including plastic fiber blankets).

b. Insulation: on top of transparency, it is an excellent insulator. According to published data (Baetensa et al., 2010), the thermal performance of a 70 mm nanogel-filled VIP is better than a 270 mm-thick hollow wall. The insulation values of a 15 mm nanogel-filled polycarbonate sheet are higher than any double glazing of similar thickness.

c. Lightness: aerogel can be three times as heavy as air (Rubin and Lampert, 1982), which means that the panels that employ this material for insulation are very lightweight even though the density of the aerogel that we use as insulation is 50-150kg/m³.

d. Versatility: monolithic aerogel can be shaped as required. Being a nano-sized filling, granular nanogel can be used to fill any chamber.

Let us note that although there are many prototypes and patents (Gibson, 2009; Bartley-Cho, 2006) for the transparent and translucent high energy efficient aerogel-implemented façade panels under study, they are extremely difficult to analyze because there is not a lot of information available and it is not easy to get physical samples of these panels. For this reason, this research has focused on commercial products that are on the market. Most of these products use granular silica aerogel (nanogel).

In the following, we analyze these translucent and transparent panels, setting out their strengths and weaknesses and our findings as a result of this study.

2.1 Translucent systems

In this type of systems we have analyzed systems composed of granular silica gel-filled polycarbonate, reinforced polyester and double glazed vacuum insulated panels.

Nanogel-filled cellular polycarbonate panels are the most widespread system on the market. They have the following strengths and weaknesses:

Strengths: Thanks to its low density 1.2g/m³, this is a very lightweight material. It has a high light transmission index ±90% (almost the transparency of methacrylate). It is a low- cost material for immediate use. And, at the competitiveness level, it is the least expensive envelope assembly.

Weaknesses: Durability is low. Most commercial brands guarantee their polycarbonate panels for only 10 years (as of when they start to deteriorate), whereas nanogel has a very high durability. These panels are very lightweight but very fragile to impact. Even though nanogel is an excellent acoustical insulator, the slimness of these panels means that they have acoustic shortcomings.

No more than two types of reinforced polyester panels are commercialized despite the potential of this material. They have the following strengths and weaknesses:

Strengths: Good mechanical properties: glass fiber reinforced polyester resin core composites offer excellent flexibility, compressibility and impact resistance. Good malleability: they could be shaped according to design needs but no existing system offers this option. Durability is good, as there are methods to lengthen the material's service life considerably (twice that of polycarbonate), like gelcoat coatings or protective solutions with an outer layer composed of a flexible "glass blanket".

Weaknesses: There is no self-supporting (structural) panel that is standardized and commercialized worldwide. Existing systems have design faults, as they include internal aluminum carpentry or substructures, whereas there is, thanks to the characteristics of reinforced polyester, potential for manufacturing a self-supporting panel (as in the case of single-hull pleasure boats). It is also questionable ecologically, as the polyester is reinforced with glass fiber, which has detracted from its use in building. However, this could change with the advent of new plastic and organic fibers and resins. Economically speaking, reinforced polyester manufacturing systems are very expensive, because either processes are not industrialized or, on the other hand, they rather technology intensive like, for example, pultrusion.

Double glazed vacuum insulated panels (VIP) are still at the prototype stage. Although research and prototypes abound (HILIT+ y ZAE BAYERN, for example) (Fricke, 2005), there are only a couple of commercial brands:

Strengths: Thanks to the combination of vacuum and aerogel insulation (both monolithic and granular), they provide the slimmest and best insulation system in the building world (0.5W/m^{2}K). Transparency levels for some prototypes using monolithic aerogel are as high as 85% for thicknesses of 15 to 20 mm. Additionally, the service life of the glazing and the gel is very similar.

Weaknesses: Product of a combination of vacuum core and double glazing, this component is fragile, especially prone to impact-induced breakages. The high cost of molding glass into complex geometries rules out its use as a free form system. It is a system that depends on substructures and other components for use (Figure 1).

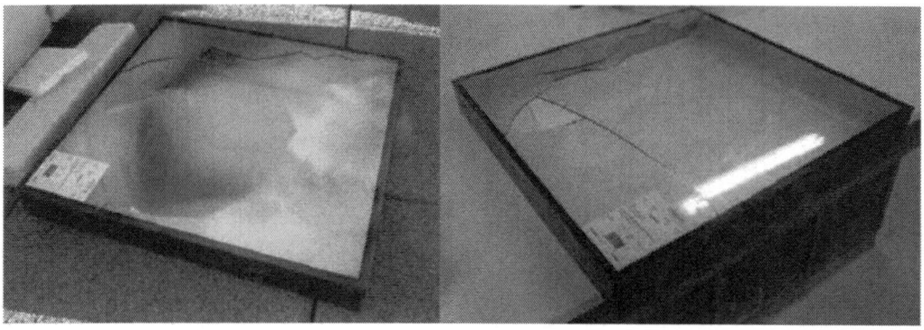

Figure. 1. Brittle fracture of panels tested in the laboratory of the Department of Construction and Technology in Architecture, School of Architecture. (UPM.) E.T.S.A.M.

Findings: After a comparative analysis of over one hundred and forty seven (147) commercial products, and the detailed evaluation of the best eight (8) (Figure 2), we can confirm that fiber reinforced polyester resin panels perform better than the best polycarbonate panels. But these improvements are unable to offset their high production and environmental costs, generating commercially uncompetitive products.

These panels have two unexploited design lines, such as adaptation to new less harmful natural cellulose resins and fibers, the design of insulation for variable geometry translucent skins, or structural improvement for use as a self-supporting component.

As regards energy efficiency, these products offer more energy-saving performance than a 27cm thick traditional wall. With only 7 cm, they have an U value of 0.28W/m^{2}K, which amounts to 7% better energy performance compared to a traditional wall and with 4% less thickness.

Looking at double glazed VIP panels; the data indicate that, although still at the prototype stage, panels like these are the best commercial solution, as they offer the best thermal and acoustical insulation performance and optimal light transmission

At the acoustical and thermal level, the VIP panel is the best of the envelopes examined, as a 15 mm panel insulates equally as well as a traditional mass wall in terms of energy expenditure (saving). We find that, with a thickness of just 60 mm, this product improves the energy efficiency performance of a 271.5 mm cavity wall.

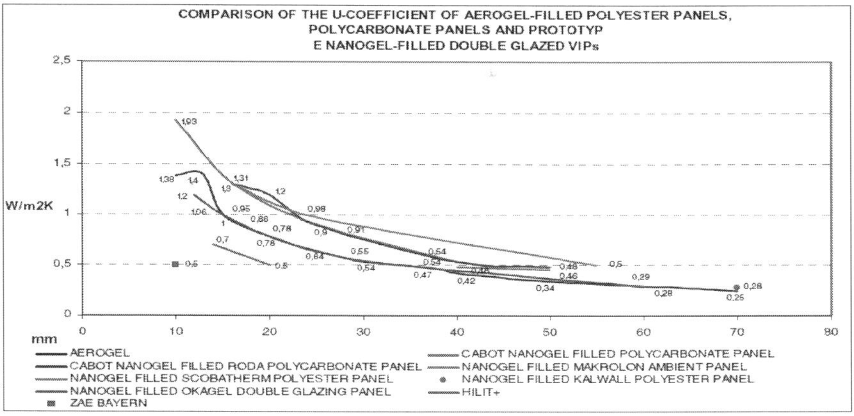

Figure. 2. Comparison of the U-value of the most significant commercial systems and prototypes implemented with an aerogel core existing on the market.

2.2 Transparent systems

All panels implemented with aerogel instead of nanogel are transparent. They have a high solar transmittance and low U value. At present all these systems are non-commercial prototypes, about which little is known. Noteworthy are two aerogel-insulated double- glazed vacuum insulation panels (VIP):

a. 4-13.5-4 / 21.5 mm double-glazed vacuum panels filled with monolithic aerogel with a pressure of 100hPa1 in the aerogel chamber. The heat transfer coefficient Ug has a U value of 0.7 W/m^{2}K for 14 mm and 0.5 W/m^{2}K for 20 mm compared to the 1.2W/m^{2}K offered by 24 mm commercial nanogel-filled double glazing VIP. This almost doubles the insulation performance of the best commercial translucent panel. Light transmission depends on the angle of incidence, but varies from 64.7 to 87.5%. The sound attenuation index is 33dB for a panel thickness of 23 mm and noise reduction is expected to be improved to 37 dB. The energy saving compared with a dwelling that is glazed with gas-insulated triple glazing (argon and krypton) is from 10% to 20% greater.

b. mm double-glazed vacuum insulation panels with aerogel spacers inside the core (unlikely to be commercialized for another two or three years). The heat transfer coefficient Ug for 10 mm panels has a U value of 0.5 W/m^2K. This is the best of all the panels studied so far, where light transmission is equal to glass.

Findings: From the analysis of the transparent panels, VIPs unquestionably come the closest to what we are looking for in this research. The only arguments against VIPs are that they are at the prototype stage. This means that they are not on the market, nor have they been tested, approved or industrialized. All this has an impact on cost. Also being the product of evacuating double glazing panels, VIPs are very fragile. Versatility is limited because, owing to the panel generation process, the maximum dimensions to date are 55 x 55 cm.

From the materials technology analysis, we find that transparent monolithic aerogel- insulated VIPs are the material that best conforms to the goals of transparency, insulation and lightness, provided that we accept that the panels are fragile, non-commercialized prototypes and are not very versatile in terms of size.

3. EXPERIMENTAL STUDY

Following up the results of the theoretical study outlined in Section 2, we now compare these findings with the results of an empirical experiment and computer-simulations of the real commercial panels to which we had access.

3.1 Computer simulation

Existing programs in the market for calculating the energy efficiency are very focused in the pursuit of thermal comfort inside the building. Therefore, we have had to adapt this program's calculating tools to finish getting results as realistic as possible.

On one side we have had modeled the environment and external environmental conditions of the test site: For this we have used the additional tools of Autodesk calculation programs, such as the Solar Tool (Figure 3) that parameterized the values of sun exposure during the test period, depending on the latitude and longitude of the testing place, or the

Weather Tool that transforms all the atmospheric data (wind, temperature, rainfall, percentage of clouds, etc.) in to numerical parameters for use them in the thermal computation.

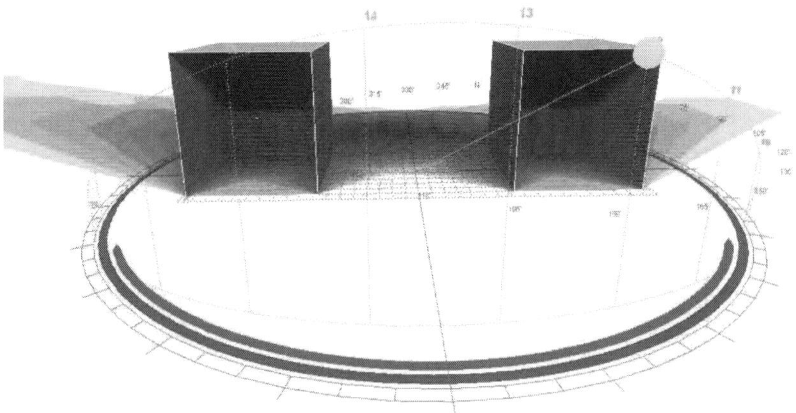

Figure. 3. Computer simulation: Solar data of March 14, 2010 at 12:00 in Madrid. (Latitude: 40 ° 23' North, Longitude: 4 ° 01' West)

On the other hand we has worked with a small-scale models, modelling the real test box, that is a cube of 60x60x60 cm, when such programs are more used to calculate large volumes built, with the adjustments that this has brought. We also have had to model all the panels systems that we tested, with the materials that are composed, in order to include them into our material library.

This whole process of modeling has been done trying to rely on independent test data. We do not want to use the data provided by trademarks, because after all, are the data that we want to verify objectively.

These programs are designed for regulatory compliance (UK and USA, among other). So the programs are oriented to guarantee thermal comfort, through adding contributions from outside air, and even turning on some air conditioning systems without the programmer's requests. Therefore, in order to get actual results, we had to overcome programming obstacles, disabling predetermined functions in some of the programs.

We used two powerful programs for the computer calculation: DesignBuilder and Ecotect. Although we have worked with both programs at the calculation, at the end we decide using DesignBuilder at the final testing, mainly due to the support the Ministry of Energy of the United States offered to the program and because the data that offered by this program correspond more to those data obtained with trials.

Because of the shortage of information about aerogel and the impossibility of acquiring a sample, we decided to use the Design Builder program to conduct a trial by computer simulation under the same environmental conditions as the empirical trials run on the other panels. Figure 2 describes the behavior of a 25 mm aerogel sheet. We find that the test space has a uniform inside temperature of between 18 and 37 ºC.

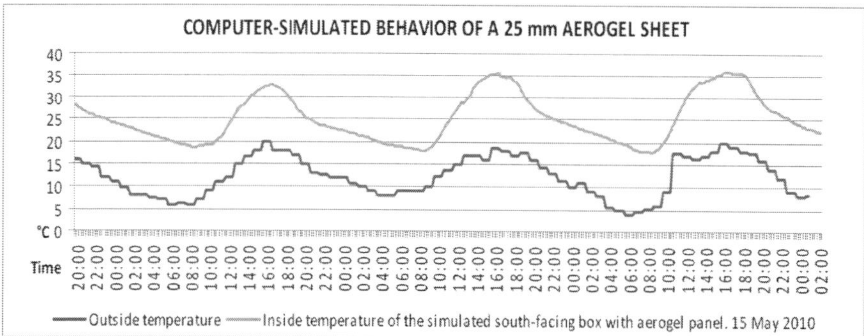

Figure. 4. Computer simulation for a 25mm thick sheet of silica aerogel with density 50- 150kg/m³ over 78 hours.

Figure 4 describes the behavior of a 25 mm Aerogel panel. We checked that the temperature inside the test space is evenly maintained between 18 and 37 °C. The curves at the loss periods are stretched, which means that this is a good thermal insulation, as it loses energy at a very long term and gradually.

3.2 Empirical trials

The trials are designed especially to examine the energy performance of the material. These trials were run at the Department of Building and

Architectural Technology of the UPM's School of Architecture on boxes with an inner volume of 60 x 60 x 60 cm, insulated with 20 cm of expanded polyurethane. One of the box faces is left open by way of a window. The study elements are placed in this opening using a specially insulated frame. The trial involves exposing two such boxes to a real outside environment to study their behavior. The two boxes have two different windows: one is fitted with 6+8+6 double glazing with known properties as a contrast element and the other is fitted with the panel that we want to study. Data loggers are placed inside each box for monitoring purposes to measure and compare their inside temperature. The boxes are also fitted with a thermal sensor on the outside to capture the temperature to which they are exposed. The boxes are set in a south-facing position as this is the sunniest exposure (Figure 5).

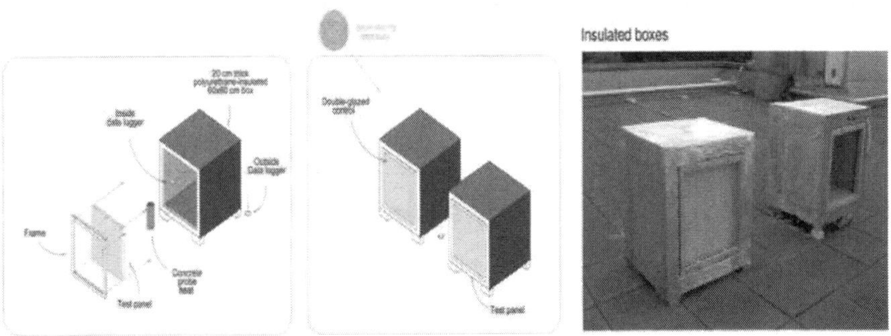

Figure. 5. Energy performance testing of the UPM system based on the hot box method

We ran twenty-eight (28) temperature-measuring trials using this system, and compared the performance of different thicknesses of commercial panels with 6+8+6 double glazing. Four (4) of these panels deserve a special mention.

3.2.1 Trial-1: 16mm nanogel-filled Cabot Lexan Thermoclear polycarbonate sheets (PC sheets)

Results: There is on average a 2.5 ºC improvement in thermal properties over the commercial Climalit double glazing (which is 5 mm thicker than the panel) at night, and it insulates almost 6ºC more than double glazing exposed direct solar radiation (Figure 6).

The fast enough rise of the curve at time of the thermal capture (day) is very similar in the two samples tested and compared (in the study panel and in the double glazing window sample probe), which means that their qualities are very similar solar collection, but there is a delay at the temperature increase with a difference of 24 minutes for the polycarbonate panel filled with nanogel, between one rise in the temperature and the other (between days). This confirms the market values in which, although showing a similar response of the double glazing, the polycarbonate panels filled with nanogel of similar thickness, usually have a small improvement if we compared them with a double glazing window.

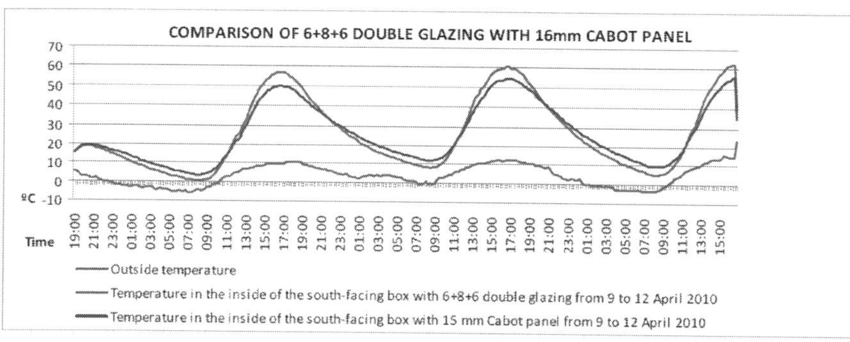

Figure. 6. Trial 1. Sample tested: (600x600mm of side) olicarbonate Panel of 16mm thick filled with nanogel. The test was made from the 9th (19:00 h) to 12 (15:00 h) in April 2010. At the E.T.S.A.M. School of Architecture & School of Computing (UPM) Spain

With the night loss, the difference between items is growing. The loss curve of the plate that is filled with nanogel is much more horizontal (which shows a U-value lower) than the loss curve of the sample window, leading to reduce the energy loss of the panel, and keeping the heat within the interior volume for more than 75 minuts and with a temperature that never is down the 10 °C although the outside temperatures was in the order of -4 °C.

These results support that a 16 mm polycarbonate panel filled with nanogel offers some benefits (although similar) than those offered by the double glazing of 20 mm (6+8+6).

3.2.2 Trial-2: 25mm nanogel-filled Cabot Lexan Thermoclear (triple-wall) polycarbonate sheets (PC sheets)

Results: There is on average a 2 to 3 ºC improvement in thermal properties over the commercial Climalit double glazing (which is 5 mm thinner than the panel) at night, and it insulates 15 to 20 ºC more than double glazing exposed to direct solar radiation (Figure 7).

The solar collection curves are very different in this graph (figure 7). While in the double glazing, the rise of the curves is very fast and in 150 minutes (two hours) the curve has risen to nearly its maximum inside temperature, the panel filled with nanogel creates a flatter curve that marks a more homogeneous rise, and it takes 300 minutes (five hours) to reach its maximum inside value. Furthermore, there is a heating delay of 120 minutes for the box where we are testing the polycarbonate sheet filled with nanogel. All of this clearly shows the best performance of this panel versus the double glazing in daytime solar collection.

The loss curves at night make a similar trend, a difference of 360 minutes, six-hour delay between the thermal losses of 25 mm panel filled with nanogel and the double glazing window. A flat graph denotes a small U-value. Also noteworthy is that the graph of the polycarbonate panel filled with nanogel oscillates in a strip of between 25 ºC and 40 ºC as the maximum internal temperature inside the box, whereas the graph of the double glazing is between 55 ºC and 10 ºC.

These data show that increasing the insulation thickness leads a heat improvement of the system that is almost directly proportional to the thickness. With a increase of the 36% of the thickness, we obtained a thermal efficiency of 33% (20 °C improvement) in the solar collection and (5 ºC improvement) in night losses.

3.2.3 Trial-3: Bayer Makrolon Ambient S2S-25 sheet. 25 mm nanogel-filled twin-wall polycarbonate panel

Results: Behavior is very uniform. We get a 3 ºC improvement in thermal properties over the commercial Climalit double glazing, which is 5 mm thinner, at night, and it insulates 5 ºC more than double glazing exposed to direct solar radiation (Figure 8)

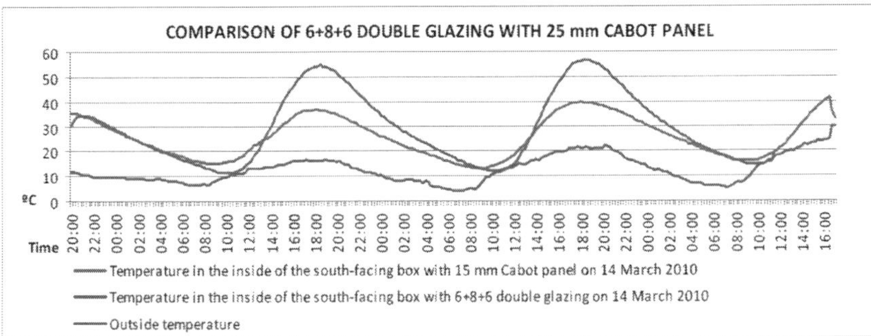

Figure. 7. Trial 2. Sample tested: (600x600mm of side) Policarbonate Panel of 25mm thick filled with nanogel. The test was made from the 14th (20:00 h) to 17 (16:00 h) in May 2010. At the E.T.S.A.M. School of Architecture & School of Computing (UPM) Spain

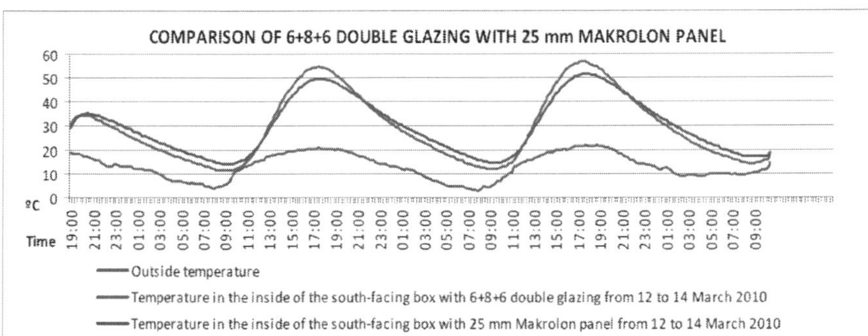

Figure. 8. Trial 3. Sample tested: (600x600mm of side) Policarbonate Panel of 25mm thick filled with nanogel. The test was made from the 12th (20:00 h) to 14 (16:00 h) in March 2010. At the E.T.S.A.M. School of Architecture & School of Computing (UPM) Spain

The solar collection curves (day) are very similar in both the study panel as in the double glazing sample and ascend quickly. Even though that thermal uptake quality are very similar in both elements, there is a delay difference of the temperature increase inside the box where the polycarbonate panel is tested of 45 minutes over the double glazing window.

At night loss the panel filled with nanogel is more flat from the start, marking quick differences with the double glazing. This trend lasts all night which denotes a small U-value and keeps the heat within the interior volume of the box over 80 minutes.

This confirms the market values, which showed that despite having the same thickness of panel (25 mm) and use the same Cabot nanogel, there are big difference between the Makrolon panel and the Lexan Thermoclear panel.

3.2.4 Trial-4: 70 mm Okagel Okalux VIP Panel: Nanogel-filled vacuum insulation panel

Results: Temperature is homogeneous ranging between 17 to 32ºC, and there is an almost constant difference of from 3 to 10ºC compared with double glazing (Figure 9).

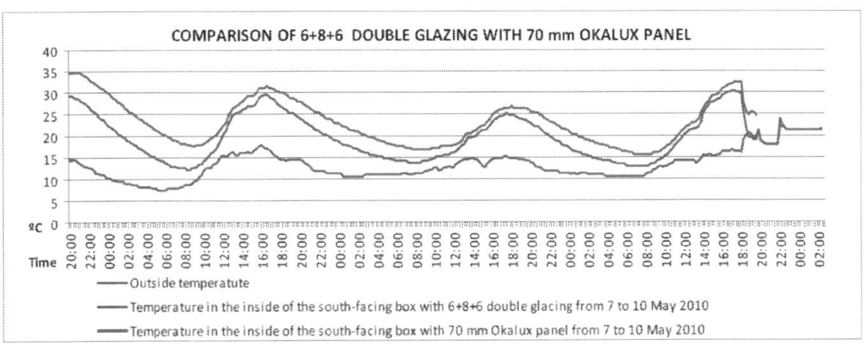

Figure. 9. Trial 4. Sample tested: (600x600mm of side) double glazing Panel/window of 70mm thick filled with nanogel. The test was made from the 7th (20:00 h) to 10 (16:00 h) in March 2010. At the E.T.S.A.M. School of Architecture & School of Computing (UPM) Spain

These four trials were evaluated and compared with the computer-simulated aerogel data (Figure 10) and data from the theoretical study. We found that, like the data output by the theoretical study, the real trials suggest that the materials behavior is suitable for designing the new envelope system. The very flat loss curves in the plot describe a very low U value. In terms of capture, there is a thermal difference of almost 30ºC between the Okagel (VIP) panel and the worst of the tested panels. The difference between Okagel and the best- performing panel is almost 10 ºC in terms of loss and capture. We have confirmed the experimental datum that likens the behavior of the Okagel panel to that of the computer-simulated aerogel.

From our computer-simulated experimental study, the data on organic aerogels supplied by the CSIC and the University of Barcelona, and the data from trials run at the University of Denmark on envelopes implemented with monolithic silica gels and the empirical trials conducted in this research, we arrive at the following conclusions:

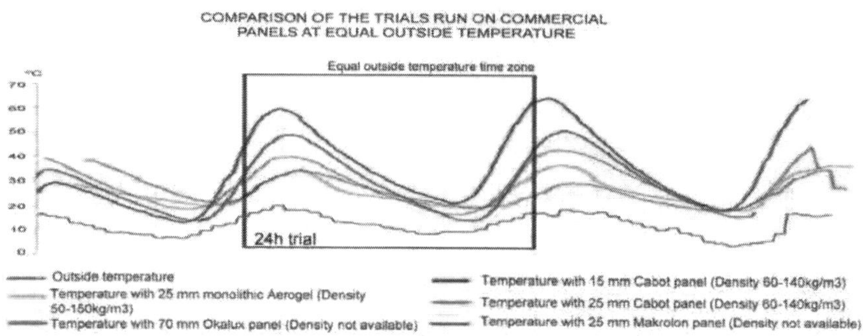

Figure. 10. Comparison of empirical data of commercial systems with computer simulation of 25 mm thick sheet of silica aerogel with density 50-150kg/m^3 over 78 hours (temperatures inside the test boxes)

The best-performing panel from the energy saving viewpoint is 70 mm Okalux that has thermal differences with respect to the other panels ranging from 5 to 20ºC. Also striking is the disparity in the results of the 25 mm and 15 mm Cabot panels with thermal differences of 15ºC.

Although the specific temperatures and factors on each test day differed from one trial to another, the 70 mm nanogel-filled Okalux VIP panel performed similarly to the 25 mm aerogel sheet.

These are key data that are useful for designing a new lightweight, slim, high energy efficient, light-transmitting envelope system, providing for seamless, free-form designs for use in architectural projects.

4 FREE-FORM, HIGH ENERGY PERFORMANCE, TRANSPARENT ENVELOPE SYSTEM (F^2TE^3) DESIGN

The proposed component has better properties than double glazing VIP and reinforced polyester panels (Okalux and Kallwall) separately, as it combines the properties of the two to generate a new system and solve the problems specified earlier.

This section will present the "Proposal for a Free-Form, Transparent, Energy Efficient Envelope System (F^2TE^3)", and will explains the details of the new system, materials of construction, its size and overall value. Also shows the way to manufacture our system and the On-site assembly of the new system. We make a special reference of how the F^2TE^3 response to the weaknesses of VIP systems. Finally, we study the tests to determine the level of energy efficiency by calculation by computer.

4.1 Proposal for a Free-Form, Transparent, Energy Efficient Envelope System (F^2TE^3)

We propose a free-form design envelope system fabricated with cellulose fibers and polyester resin (or acrylic-based organic resin), and a vacuum core insulated with monolithic aerogel at a pressure of 1hPa. Being a self-supporting component, the system can perform structural functions, and seams between panels are concealed by an outer coating applied in situ (Figure 11).

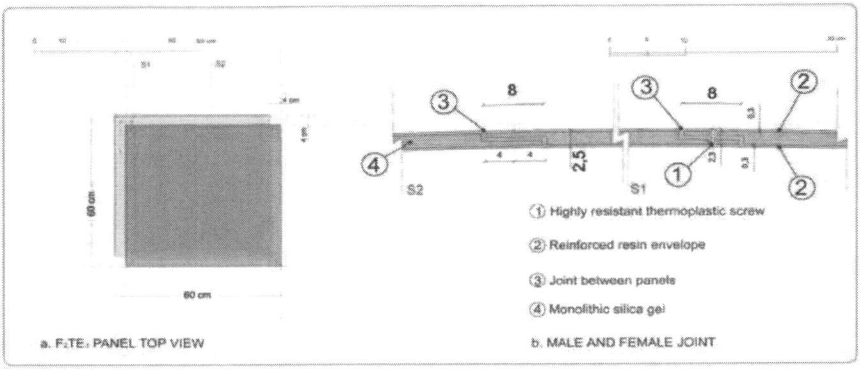

Figure. 11. F^2TE^3 Envelope Profile Design

Materials

The optimal materials are: (a) a natural cellulose fiber-reinforced epoxy resin matrix with similar performance to E-type glass fiber, with an outer gelcoat coating to protect it from external agents for the outer skin, on which the panel's resistance, protection and variable geometry depends, and (b) a thermal/acoustical insulation component composed of a monolithic silica gel-filled vacuum chamber.

Dimensions

Based on tests run at the University of Denmark, we have to take into account the sol-gel process drying times required for the monolithic silica gel to generate a crystalline structure, the percentage of breakages due to size and, above all, the fact existing autoclaves are able to generate monolithic gel pieces no larger than 55 x 55 cm. The panel sizes will be 60 x 60 cm (length/width), and panel thickness will depend on whether the structure is self-supporting or a simple envelope. For modeling purposes, however, we have studied a 25mm thick panel, composed of two sheets of 3 mm thick reinforced resin and a vacuum core filled with monolithic silica gel (Figure 12).

The weight per unit of surface area will be from 15 to 7 kg/m2. Although dimensions could vary, the sheet width will be no greater than 600 x 600 mm and the minimum admissible flexion radius will be approx. 4000 mm.

Specifications

Light transmittance, τD65: from 59% to 85% approx. UV absorption: approx. 20%

Total energy: approx. 61%

Horizontal and vertical U-value: 0.50 W/m2 K

Thermal conductivity coefficient: 0.01 W/m K, estimated Possible heat-and humidity-induced dilation: 3 mm/m approx. Operating temperature: -70 to +80 °C

Weighted sound reduction value: estimated at 26-45 dB

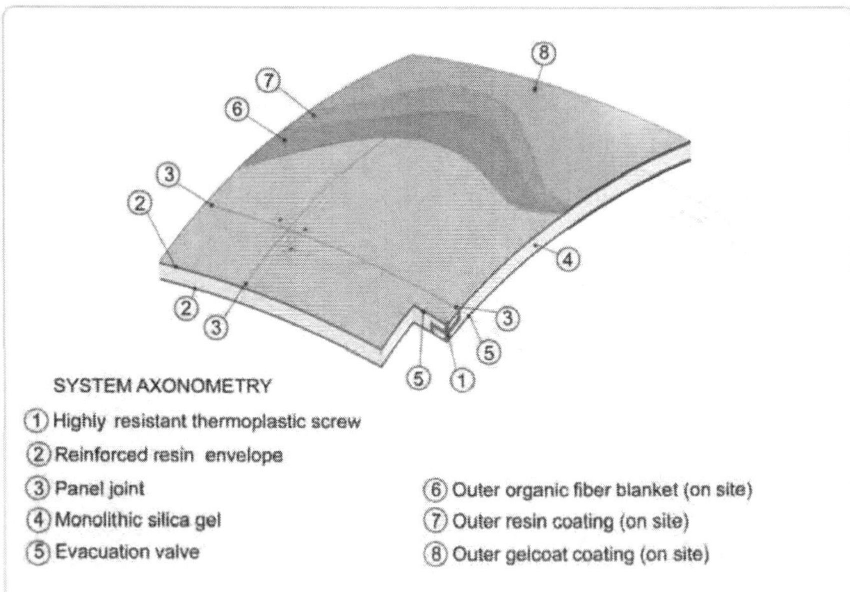

Figure. 12. F^2TE^3 system axonometry

4.2 System specifications

The F^2TE^3 is composed of the dry-seal connection of previously designed male and female edged panels (two female sides and two male sides on each panel that fit together seamlessly). Once the construction is in place, it is given an outer coating of fibers and resins and finally a gelcoat coating to protect the assembly from external agents.

Manufacturing and on-site assembly (Figure 13)

The off-site manufacturing process is composed of the following phases:

An easy-to-use, reusable and ecological molding process. As this does not have to be a structural mold, it can be made of compressed and sanded sheeting arranged according to the panel design.

The same mold is used to generate the two panel ends and its walls.

The walls are adhered to one of the ends using resin, and the assembly is filled with silica gel, where the panel itself acts as a mold to shape the monolithic silica gel inside the chamber.

Sol-gel technology is used to generate the aerogel inside the panel using an autoclave.

The top end is adhered to the panel using resins.

The panel is evacuated to a pressure of 1hPa inside the chamber.

The panels are transported separately to the site. Scaffolding and props are then used to assemble in situ the panels making up the façade. The panels are screwed together using highly resistant transparent thermoplastic screws.

The system should be manufactured under adequate health and safety conditions.

4.3 Testing

A F^2TE^3 system with a thickness of 25 mm has been computer simulated to examine its energy-saving behavior compared with a computer-simulated aerogel envelope of the same thickness (Figure 14).

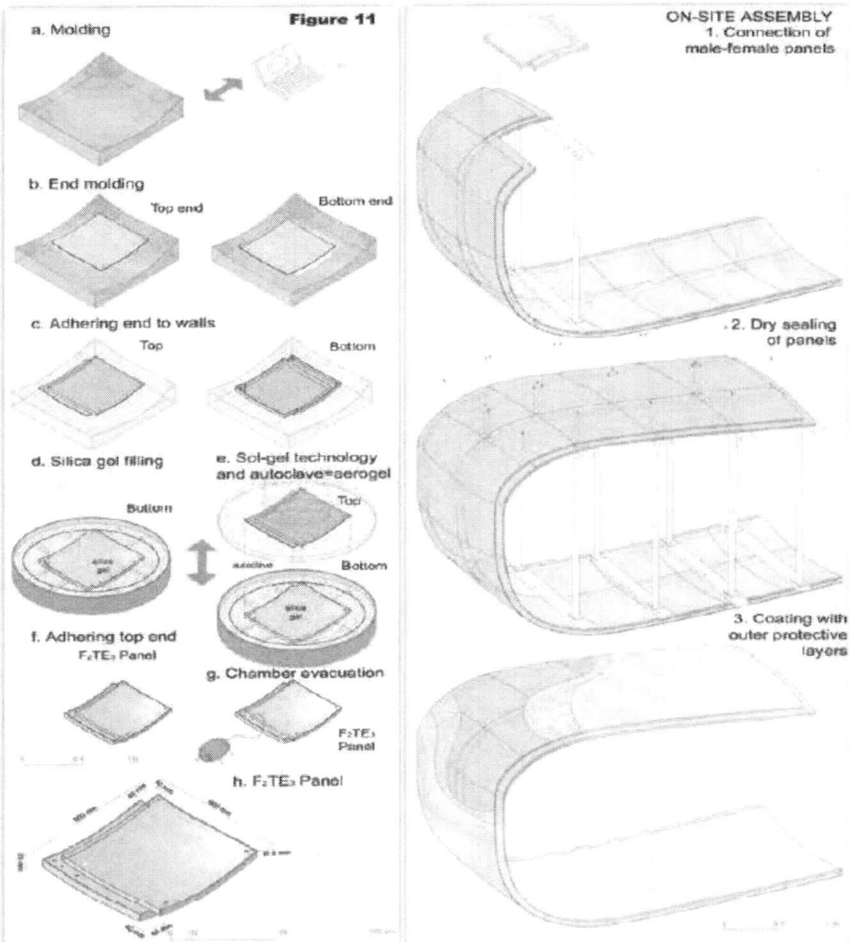

Figure. 13. Manufacturing and On-site assembly

As shown in Figure 12, although the plots are displaced, the F^2TE^3 system returns a result very close to what would be achieved with monolithic aerogel without a barrier envelope (not feasible due to aerogel hydroscopy). Even with a barrier envelope, F^2TE^3 performance almost equals aerogel in terms of heat loss, with a very similar flat curve, where the U value is very small, but results for capture are worse at over 5ºC higher.

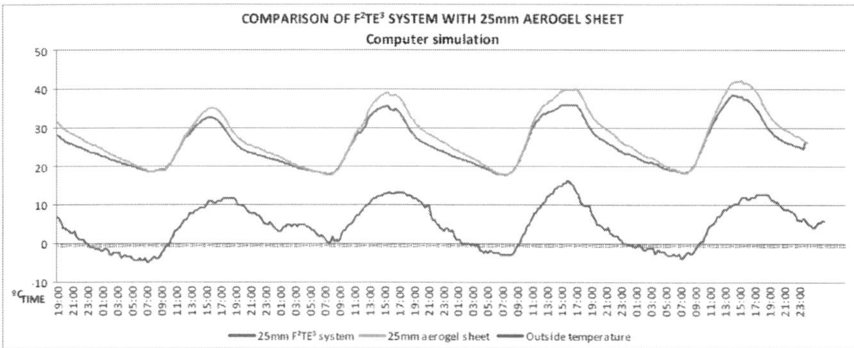

Figure. 14. Comparison of a computer simulation of a 25mm thick sheet of silica aerogel with a density of 50-150 kg/m³ with the F^2TE^3 system over 96 hours.

4.4 F^2TE^3 response to the weaknesses of VIP systems

Most studies conducted in the field of VIP elements (VIS) (Mukhopadhy et al., 2011) determine that there are four key obstacles to the use of these systems in the building industry: a) fragility of the outer skin, b) thermal bridging at panel edges and joints, c) vacuum chamber moisture permeation, and d) price. On top of these weaknesses, there is the added obstacle of a single element having to provide transparency and energy efficiency, plus a new demand from vanguard architecture for generating free forms by means of monocoque systems.

Theoretically, the new F^2TE^3 overcomes all the above objections raised against VIPs, offering the added values of transparency and free-form design:

a. Fragility of the outer skin: As the new system has to be a transparent and self- supporting structure, we have replaced the fragile outer barrier elements, such as a metallized polymer multilayer outer skin, double glazing and thin stainless steel sheets, with an element composed of highly resistant reinforced fibre resin.

b. Thermal bridge at panel edges and joints: Commercial panels are not designed to be assembled to form seamless elements. The new system that we propose is purposely designed to generate monocoque elements through a system of male and female panels. This system eliminates the thermal bridge between the panels (see Figure 15).

c. Moisture in the vacuum chamber: One of the key problems of VIP systems is condensation forming inside the chamber. Although some studies estimate that the hydroscopic silica smoke is capable of absorbing most of the minimal amount of condensation that forms without compromising its load-bearing or thermal capacities, the fact is that humidity inside the chamber ends up turning the monolithic aerogel opaque. F^2TE^3 solves this problem by modifying the barrier thickness. The outside face is 3 mm thick compared with the inside face that should have thicknesses ranging from at least 6 to 8 mm to prevent condensation forming inside the chamber. We used the COAG's memorias2 program to plot the condensation graph of the new envelope, and the resulting sections are shown in Figure 16.

F^2TE^3 has a water vapour permeability value of 0,3 g/m²d (for 3mm thickness) less than some commercial VIP panels (0,5 g/m²d)

Price: VIP systems are now much higher cost than traditional building systems (approximately 200 €/m²). However, Technical University of Denmark (DTU) studies (Jensen et al., 2005) show that the mass produced product, using a fully automatic 10 m³ autoclave (as the sol-gel process is the most expensive part of the fabrication of this system), could be commercialized for less price.

For the purposes of our calculations, we will take a 150 m² detached dwelling and assume that the F^2TE^3 system costs 300 €/m² (If we use an conventional autoclave of 600 ml capacity (Parr Instrument Company, Moline, IL, USA) from the Air Glass Laboratory). On the other hand, the mean cost of a 6-12-6 double glazing aluminium window with a thermal break is 360 €/m², whereas a plastered, double hollow brick lined, 4 cm thermally insulated, 6 inch rough brick conventional wall with single-layer finish costs on average 85 €/m². Assuming that the mean percentage of glazing is 45%, the current cost of a conventional envelope including the façade glazing is 162 €/m².

The surface area of the envelope for a 150 m² dwelling is 375 m². The cost of building a conventional wall will be 60,750 €, whereas F^2TE^3 costs 112,500 €, meaning that F2TE3 costs just over double the traditional system at this stage.

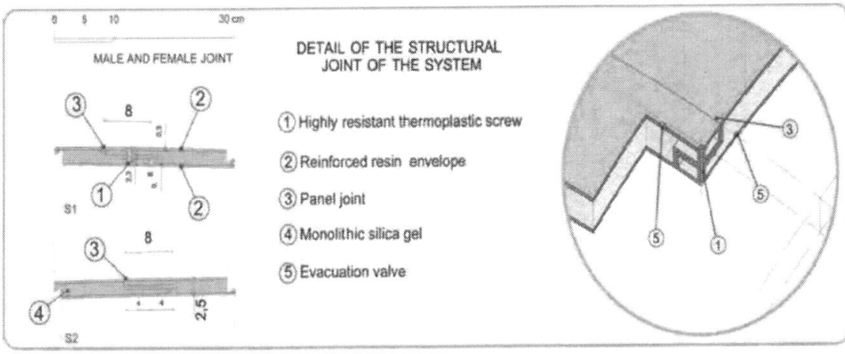

Figure. 15. Detail of the male and female joint of the F2TE3 system.

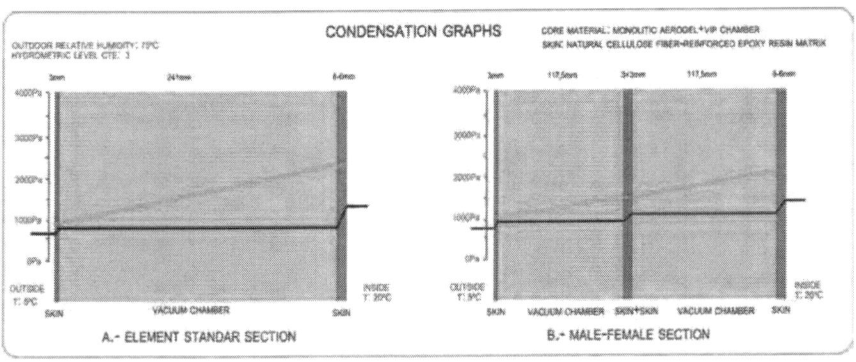

Figure. 16. Condensation graphs for a standard section and a male female section of the new F2TE3 system.

Note, however, that the net internal area of the dwelling built using the F^2TE^3 system is 148 m^2 (at 2.5 cm thick the envelope only takes up 0.833% of the gross external area), compared with 135 m^2 for the traditional system (at 30cm thick it takes up 9.7% of the gross external area). Therefore, developers using the new F^2TE^3 system can commercialize an additional net internal area of 13 m^2. Assuming that the final sale price per square metre of net internal area is 4,000 €, there will an additional gain of 53,560 €. This, deducted from 51,750 €, which is the result of subtracting the price of the conventional building skin from the cost of the proposed envelope, results in a net profit of 1,810 € for developers using F^2TE^3.

Clearly, this calculation does not take into account either the potential saving thanks to the use of the new system as a structural element (which could amount to a 20% saving over the total cost of executing the building works) or the potential time and labour saving using this new building system compared with the traditional envelope.

According to DIN standard 4701, "there would be an annual saving of from 0.9-1.3 l of oil or 1.0-1.5 m^3 of gas per m^2 of naturally lit surface area, leading to U-value reductions of 0.1 W/m^2K". The F^2TE^3 surface area would be the sum of the walled areas (168 m^2) and glazed areas (206 m^2). Accordingly, with a U-value of 0.35 W/m^2K, compared with 0.44 W/m^2K for the conventional wall and 1.2 W/m^2K for the glazing, the proposed system achieves a saving of 2083 l of oil (0.7€/l) and 2403m^3 of gas (0,55€/ m^3). This is equivalent to an annual saving of 1458€ in oil or 1322€ in gas for the consumer and a CO2 reduction of approximately 4.9 kg per year.

Transparency, energy performance and variable geometry: Of the three types of translucent and transparent insulation now on the market (plastic fibres, gas and aerogel), we found that the best core material for the new F^2TE^3 system is aerogel, (Ismail, 2008; Wong et al., 2007) as it offers:

Transparency: aerogel compares with glass in transparency (87.6%)

Insulation: aerogel is an excellent insulator. According to published data, [4] the thermal performance of a 70 mm nanogel-filled VIP panel is better than a 270 mm cavity filled wall.

Core material: thanks to its porosity and friction joint sealing, silica smoke is the material that many studies (Musgrave, 2009) recommend as VIP core material, provided that there is an outer protective skin. This also has a barrier function in order to maintain the vacuum of the panel.

Lightness: aerogel is only three times heavier than air, and although the density of the aerogel that we use as insulation is higher (50-150kg/m^3) remains very light.

Versatility: monolithic aerogel can be shaped as required (Tenpierik & Cauberg, 2006).

5. CONCLUSION

F^2TE^3 is a slim façade system that provides high energy efficiency, with a seamless surface, providing for variable geometry and the option of building self-supporting structures into the same transparent system skin. The study conducted as part of this research has shown that the prototype F^2TE^3 system outperforms other systems existing on the market, offering added value in terms of structure, transparency and variable geometry, and overcoming the traditional weaknesses of most of the VIP panels.

Computer-simulated trials have shown it to have almost identical energy efficiency properties to monolithic aerogel systems and VIP envelopes. This system revolutionizes VIP systems, as it generates a transparent envelope but eliminates breakages due to fragility by substituting glass for a reinforced composite material. Additionally, it offers the option of generating variable geometry designs.

The prototype F^2TE^3 system outperforms the systems existing on the market by combining some of the best properties of these systems and solving their weaknesses.

ACKNOWLEDGMENT

The development of this research work has required the participation of numerous comercial companies, that have provided information and material that made this research feasible. To this end, I quote Wim Van Eynde Makrolon Ambient's Commercial Director (Belgium), Christina Beck, Director of the Office Sales and Export of Okalux (Germany) and Silvestre Shana Market Development Manager of the Cabot Corp. company (USA).

We also want to thank for their collaboration to Anna Roig researcher at the Materials Science Institute of Barcelona (ICMAB-CSIC) and the support of the ABIO research group of E.T.S.A.M., the Department of Construction and Technology in Architecture, School of Architecture (UPM) and the Languages and Systems and Software Engineering Department, School of Computing (UPM).

REFERENCES

1. Baetens, R., Jelle, B. P., Thueb, J. V., Tenpierikd, M. J., Grynninga, S., Uvsløkka, S. & Gustavsene, A. (2010). Name of paper. *Vacuum insulation panels for building applications: A review and beyond,* Vol. 42, No. issue 2, (February 2010), pp. (147-172), ISSN 0378-7788

2. Commission of the European Communities. (23.1.2008). *Moving Forward Together on Energy Efficiency* (Communication from the Commission to the Council and the European Parliament on a first assessment of National Energy Efficiency), Action Plans as required by Directive 2006/32/ec on energy end-use efficiency and energy services, 2008, COM(2008) 11 final, Brussels

3. Fricke, J. (2005). Title of conference paper, *Proceedings of 7th International Vacuum Insulation Symposium,* ISBN 3-528-14884-5, Empa, Duebendorf/Zurich, Switzerland, September 2005

4. Gibson, A. D., (Aug 2009). *Construction Panel System and Method of Manufacture Thereof,*

5. (PCT), The International Patent System, patno: wo09105468, USA

6. Heinemann, U., Weinläder, H. & Ebert, H.-P., (03 March 2009). *Energy efficient building envelopes: New materials and components* (Energy Working Group (EPA)), German Research Centre in terms of energy applications, IBS/b – 43145/00, Hamburg

7. Ismail, K.A.R. (2008). Comparison between PCM filled glass windows and absorbing gas filled windows. *Energy and Buildings,* Vol. 40, No. 5, (July 2008), pp. (710-719), ISSN: 03787788

8. Jensen, K. I., Kristiansen, F. & Schultz, J. (Eds.). (30 November 2005). *HILIT+. Highly Insulating and Light Transmitting Aerogel Glazing for Super Insulating Windows,* Department of civil engineering (BYG-DTU) Technical University of Denmark, Contract ENK6-CT-2002-00648, Lyngby, Denmark

9. Martín, L., Oriol-Ossó, J., Ricart, S., Roig, A., García, O. & Sastre, R., (2008). : Organo- modified silica Aerogels and implications for material hydrophobicity and mechanical properties. *Journal of Materials*

Chemistry, No. 18, (month and year of the edition), pp. (207 - 213), ISSN 0959-9428

10. Moner-Girona, M., Martínez, E. & Roig, A., (2002). Micromechanical properties of carbon– silica Aerogel compositesAerogeles. *Appl. Phys.*, Vol. A, No. 74, (2002), pp. (119– 122), ISSN 0947-8396

11. Mukhopadhy, P., Kumaran, K., Ping, F. & Normandin, N. (2011). Use of Vacuum Insulation Panel in Building Envelope Construction: Advantages and Challenges, *Proceedings of 13th Canadian Conference on Building Science and Technology*, N R C C - 5 3 9 4, Winnipeg, Manitoba, May, 2011

12. Musgrave, D. (2009). Structural Vacuum Insulation panels, *Proceedings of International Vacuum Insulation Symposium (IVIS-09)*, ISBN, London, Uk, September 2009

13. Rubin, M. & Lampert M., (1982). Transparent Silica Aerogel for Window Insulation. *Solar Energy Materials*, No. 7, (September 1982), pp. (393-400), ISSN 0165-1633 Irregular

14. Simmler, H., Brunner, S., (EMPA) Heinemann, U., Schwab, H., (ZAE-Bayern) Kumaran, K., Mukhopadhyaya, P., (NRC-IRC) Quénard, D., Sallée, H., (CSTB) Noller, K., Kücükpinar-Niarchos, E., Stramm, C., (Fraunhofer IVV) Tenpierik, M., Cauberg, H., (TU Delft) & Erb, M., (Dr.Eicher+Pauli AG) (Sept. 2005). *Vacuum Insulation Panels - Study on VIP-components and Panels for Service Life Prediction of VIP in Building Applications (Subtask A-B),* (HiPTI-High Performance Thermal Insulation), IEA/ECBCS, Annex 39, Switzerland

15. Tenpierik, M. & Cauberg, H. (2006). Vacuum Insulation Panel: friend or foe?, *Proceedings of PLEA2006, The 23rd Conference on Passive and Low Energy Architecture*, ISBN 2- 940156-31-X, Geneva, Switzerland, September 2006

16. Willems, W. M. & Schild, K. (2008). Where to use vacuum insulation... and where not!, *Proceedings of 8th Nordic Symposium on Building Physics*, ISBN/ISSN: 978-87-7877- 265-7, Copenhagen, Denmark, June 2008

17. Willems, W. M. & Schild, K. (2006). The use of Vacuum Insulated Sandwiches (VIS) in building constructions, *Proceedings European*

Energy Performance of Buildings Directive at EPIC 2006 AIVC Conference, ISBN:2-86834-122-5, Lion, France, 20-22 November 2006

18. Wong, I.L., Eames, P.C., & Perera, R.S., (2007). A review of transparent insulation systems and the evaluation of payback period for building applications. *Solar Energy,* Vol. 81, No. 9, (Sep 2007), pp. (1058-1071), ISSN 0038-092X

Chapter 9

Ventilation of Apartment Buildings and Nursing Homes

Alo Mikola, Teet-Andrus Koiv, Hendrik Voll

Department of Environmental Engineering, Tallinn University of Technology, Tallinn, Estonia

ABSTRACT

The paper presents the results of the investigations into the indoor climate and energy efficiency of ventilation devices in residential buildings and nursing homes. The indoor climate studies discussed in the article were conducted in a nursing home where room ventilation is based on Meltem Air Handling Units (AHU). Similar studies have been carried out in apartment buildings where CO_2 levels in bedrooms were compared before and after renovating the ventilation with Meltem AHU and installing exhaust fans in the bathroom/WC and kitchen. Tenants evaluate the use of Meltem AHU in apartments very positively. The article presents the efficiency results of Meltem AHU with different external air temperatures conducted in apartments in real-life situations. The study shows that skilful renovation of ventilation in old apartment buildings enables to achieve good indoor climate and energy saving at the same time.

KEYWORDS

Indoor Climate, Apartment Buildings, Nursing Home, AHU, Energy Efficiency of AHU

1. INTRODUCTION

The paper presents the results of the investigations into indoor climate in nursing homes and residential buildings. In the review of ventilation in European dwellings, it was pointed out that the ventilation of residential spaces in Nordic countries is often poor [1] . As a result of refurbishing, the air tightness of the building envelope increases and as many buildings have natural ventilation, the air change rate is reduced. The performances of the ventilation systems in low-energy residential houses have been presented by Maier et al. [2] . The energy performance of heat recovery units has been investigated by Jaber et al. [3] and Laverge et al. [4] . Ventilation heat recovery in residential buildings has been investigated by Dodoo et al. [5] . A number of studies on the ventilation and indoor air quality in apartment buildings have been carried out at Tallinn University of Technology by Mikola and Koiv [6] -[9].

2. METHODOLOGY

2.1. Indoor Air Quality

In accordance with the requirements [10] in Estonia, the living area must have natural or mechanical ventilation, which guarantees the air change necessary for human activity. According to the requirements in Estonia [11], air velocity in living spaces, the volume of space per person and the content of harmful substances in the indoor air must not exceed the permitted values.

Values for assessing room temperatures are taken from Estonian standard EVS-EN 15251:2007 [10] . This standard describes indoor environmental input parameters for designing and assessing the energy performance of buildings. The temperature values in case of different classes of thermal comfort are described in Table 1. The values to assess the relative humidity of indoor air are taken from Estonian indoor climate standard EVS 839:2003 [12] . The parameters are described in Table 1.

On the valid regulative basis, the CO_2 concentration in the indoor air in Estonia is set in the standard of indoor environmental input parameters [10] and designing criteria CR 752 [13] . Danish University of Technology has studied the standard in detail and gives the air flow rate of the classes of indoor climate and the maximum norms of the CO_2 content in the indoor air [14] . These norms also correspond to the values that are given in the designing criteria of indoor climate. Due to that the present study uses the maximum values given in designing criteria CR 752 for accessing the CO_2 content in the indoor air, whereby the CO_2 content in the external air is considered to be 350 ppm. The values to assess the CO_2 concentration of indoor air are described in Table 2. The standard for determining the initial parameters of energy efficiency [10] allows short-time deviations from the required parameters of indoor climate. The permitted deviation is up to 5%.

According to the standard for determining the initial parameters of energy efficiency (EVS-EN 15251:2007), it is possible to deduce the airflow according to the measurements of CO_2concentrations in rooms where the main source of indoor air pollution is people. This method is called the metabolic CO_2 method. To assess the air change of a room by the metabolic CO_2 method it is possible to use Equation (1):

$$C = C_v + \frac{m}{L} - \left(C_v + \frac{m}{L} - C_0\right) \cdot \left(e^{-\frac{L}{V}t}\right) \tag{1}$$

where C is the indoor concentration of CO_2 at time t (g/m^3), C_v the concentration of CO_2 in the external air (g/m^3), C_0 the concentration of CO_2 in the indoor air at time 0 (g/m^3), L the volume flow rate of air entering the space (m^3/h), m the volumetric indoor emission rate of CO_2 (g/h), V the volume of the indoor space (m^3) and t is the interval since t = 0.

Using literature [15] , it is possible to express the relationship between the metabolic heat rate, body surface area and oxygen demand by Equation (2).

$$Q_{CO_2} = \frac{0.727 \cdot M \cdot RQ \cdot m^{0.425} \cdot l^{0.725}}{4.83 \cdot RQ + 16.17} \tag{2}$$

Table 1. Thermal classes of indoor climate based on Estonian standards EVS-EN 15251:2007 and EVS 839:2003.

Class of thermal comfort	Summer		Winter	
	Relative humidity, %	Room temperature, °C	Relative humidity, %	Room temperature, °C
I	30 - 70	23.5 - 25.5	25 - 45	21 - 25
II		23.0 - 26.0		20 - 25
III		22.0 - 27.0		18 - 25

Table 2. Class of the indoor climate for rooms with human activity (CR 1752) [13] .

Category	Expected percentage dissatisfied, %	CO_2 concentration at outdoor air level 350 ppm, ppm	Indoor air CO_2 concentration, ppm
I (A)	15	460	810
II (B)	20	660	1010
III (C)	30	1190	1540

where Q_{CO_2} is the human indoor emission rate of CO_2 (l/h), RQ the ratio of M O_2 and exhaled CO_2, Q_{O_2} human oxygen demand (l/h), m the weight of person (kg), l the height of person (m), M the metabolic heat rate (W/m^2).

Human CO_2 emission rates are mainly related to the surface area of the body and the metabolic heat rate. The ratio of inhaled O_2 and exhaled CO_2 is taken as a constant value in the present study (0.7 to 1.2 met). If the value of RG is not exactly determined, it can be considered 0.83 [15] . In this case the accuracy of CO_2 emissions is approximately 3%, which is a sufficient result.

The metabolic CO_2 emissions are often observed as the average value of a 24 h period. In this case the average metabolic CO_2 emission is calculated using the weighted average method. In some studies the metabolic CO_2 emissions are shown during the sleeping period. As the CO_2 emissions vary in a wide range in the day time, it would often be much more punctual to use the night time CO_2 emissions in the calculations [10] . In the present study night time CO_2 emissions are used in the case of apartment buildings. In this study adult CO_2 emission is 13 l/h and children's CO_2emission is 6.5

l/h during the night time. Metabolic CO_2 emissions in different studies are shown inTable 3.

2.2. Ventilation Heat Recovery

Member States of the European Union are required to implement energy efficiency measures for buildings under the Energy Performance of Building Directive [16] . With new building codes, EU member states intend to reduce the total energy consumption in buildings by making the buildings well-insulated and tighter. Due to the tighter and well-insulated constructions, the ventilation loads constitute a growing part of the heating demand between 30% - 60% for new and retrofitted buildings [17] . As the ventilation heating demand plays an important role in the building's total heating demand, the heat exhaust air heat recovery is inevitable.

From the exhaust air it is possible to recover only sensible heat or both sensible and latent heat [18] [19] . Energy is used to cover the heat losses due to ventilation air and to move the ventilation air for mechanical ventilation. The ventilation system also influences the air infiltration through the building envelope.

The temperature efficiency, which does not include latent heat transfer, was used to quantify the effect of heat recovery. The temperature ratio (efficiency) η_{temp} is defined as [18] [20] :

$$\eta_{temp} = \frac{\dot{m}_{in} \cdot \left(\overline{t_{supply}} - t_{outdoor} \right)}{\dot{m}_{\min} \cdot \left(t_{exhausst} - t_{outdoor} \right)} \tag{3}$$

$t_{outdoor}$ is the outside air temperature, $t_{exhaust}$ is the temperature of the extract air, $\dot{m}_{in}$ is the inlet mass flow rateand $\dot{m}_{\min}$ is the minimum of inlet and outlet mass flows. In the case of the regenerator, a time-averaged value of the supply air temperature $\overline{t_{supply}}$ has to be used which is given by [20] :

$$\overline{t_{supply}} = \frac{1}{\tau} \int_{t=0}^{t=\tau} t_{supply}(t) \cdot \mathrm{d}t \tag{4}$$

Table 3. Metabolic CO_2 emissions in different studies [21] -[26] .

Country, study	Object of the study	Adult CO_2 emission, l/h	Infant CO_2 emission, l/h
USA, Dietz & Goodrich 1995	Schools, residential buildings	19	12
Thailand, Leephakpreeda *et al*. 2000	Schools	16.2	16.2
Ireland, Guo and ja Lewis 2007	Residential buildings	18	18
Sweden, Pavlovas 2003	Apartment buildings	12*	12*
Czech, Jokl 2000	Schools, office buildings, residential buildings	19 (1 - 1.2 met)	18 (3 - 6 yrs and 2.7 met); 19 (14 - 16 yrs and 1 - 1.2 met)
Estonia, Kalamees *et al*. 2012	Apartment buildings	13*	6.5* (until 14 yrs)

* -night-time.

t denotes the time and τ is the semi-period, which means the duration of a supply or extract process. In the case of the recuperator, the process is in a steady state [20] :

$$t_{supply}\left(t\right) = \text{constant} \tag{5}$$

2.3. Studied Buildings

Studies have been carried out for the nursing home at Alutaguse, Figure 1. This nursing home has 39 rooms for 1 - 2 persons, the ventilation of which is solved with Meltem AHU.

The nursing home has hotel type rooms, the area is 25 m^2. The rooms are designed for 2 people (Figures 2-4) but in reality 1 or 2 people live in a room. Meltem Komfort AHU with network control has been installed in the rooms. Figure 3 also shows that with the Meltem AHU polluted air is extracted from the bathroom and the supply air is given to the living room.

Meltem AHU can be installed in the wall construction, Figure 5 or on the wall.

Meltem device has an aluminium recuperative heat exchanger, two ventilators, two filter cassettes and a control device.

Meltem room AHU has the same working principles as a standard air handling unit.

A specific feature is that the device serves up to 2 rooms, so the necessary length of the duct is minimal, typically up to 2 m, and the service is located in the AHU room. Thanks to short ducts and EC motors the electrical power of the devices is up to 15 W maximum, and hence the electric energy consumption is about 120 kWh per year, which is significantly lower than with the central balancing system.

The height of the device is 409 mm, the width is 388 mm and the depth is 196 mm. It is possible to use: standard filter G4, allergy F7 or absorbent carbon filter F6.

Figure 1. Outside view of the nursing home.

Figure 2. A two-person room.

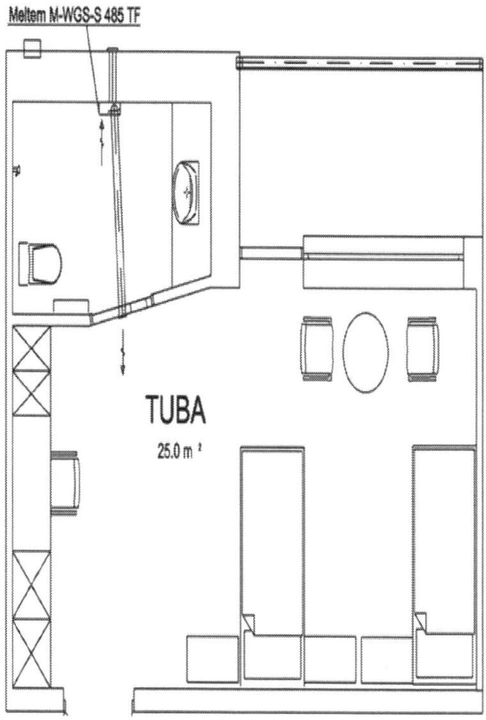

Figure 3. 2-person room solution with Meltem AHU.

Meltem AHU is available in three modifications: Standart, Standart with network control and Komfort.

According to technical documentation Standart AHU enables three flow rates: 15/30/60 m^3/h, power consumption from 3.8 to 12.5 W and noise level from 15.5 to 36 dB(A).

Research of Meltem AHU in apartment buildings (Figure 6) was mostly carried out in a 2-room apartment, Figure 7 and in 3-room apartments. The apartments were fitted with Komfort devices.

As the HVAC requirements of a nursing home and residential building are close, we can also apply the studies of Meltem AHU conducted in apartments to a nursing home.

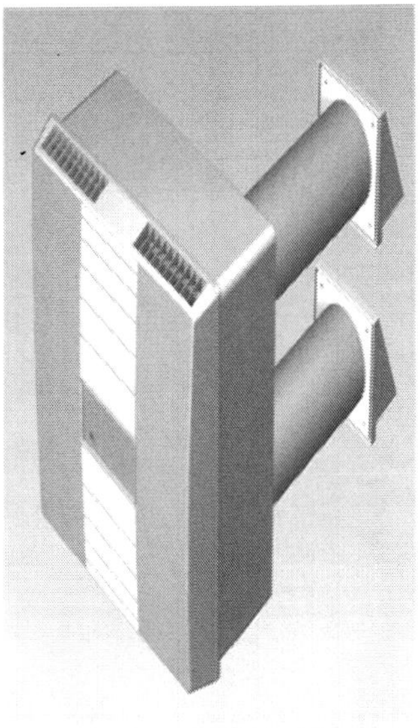

Figure 4. Meltem AHU is installed at the height of 2 m.

Figure 6 shows that one Meltem AHU is used in a 2-room apartment. In addition to AHU, an exhaust ventilator has been installed in the toilet/bathroom, which works when the lighting is switched on, and in the kitchen when required. The fans have been programmed to run up to 10 minutes after the light is turned off. After that the ventilation works as natural. The exhaust air ventilator is shown in Figure 8.

As the ventilator capacity is 64 m^3/h, this ensures good indoor climate in the room while being used. If the ventilator is not in use, the necessary air change is ensured by natural ventilation.

3. RESULTS

3.1. Measurements of Indoor Climate

3.1.1. The Nursing Home

Indoor climate studies in the nursing home were carried out in 4 rooms, 2 of them were single rooms and 2 double rooms. The recording results of

CO_2 concentration and relative humidity in different rooms of the nursing home are presented in Figures 9-11.

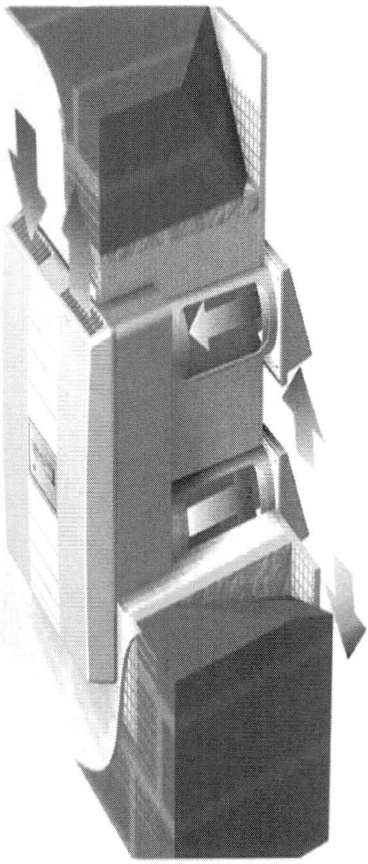

Figure 5. The operation principle of the unit Meltem.

Figure 9 shows that the relative humidity does not exceed the permissible limits of residential premises.

Economic calculations indicate that the construction costs of the ventilation on the bases of Meltem AHU and a centralised system in the nursing home are almost equal. At the same time the power consumption is lower and Meltem device allows flexible use, so the maintenance costs are significantly lower. Due to the lack of metallic ducts the number of light negative ions in the room air is also higher.

The results of the measurements conducted in different rooms show us that room air handling unit Meltem with heat recovery ensures good indoor climate in the nursing home. At the same time rational organization of ventilation by Meltem significantly contributes to energy saving compared with balance ventilation.

We see that the carbon dioxide concentration in rooms was less than 1000 ppm, which can be considered a good result, and in the majority of the rooms less than 700 ppm, which can be considered a very good result.

Figure 6. Facade of the house.

3.1.2. Apartment Buildings

In apartment buildings studies of CO_2 concentration, relative humidity and temperature were carried out after renovating the ventilation. CO_2 level after the installation of Meltem AHU is shown in Figure 12 and Figure 13.

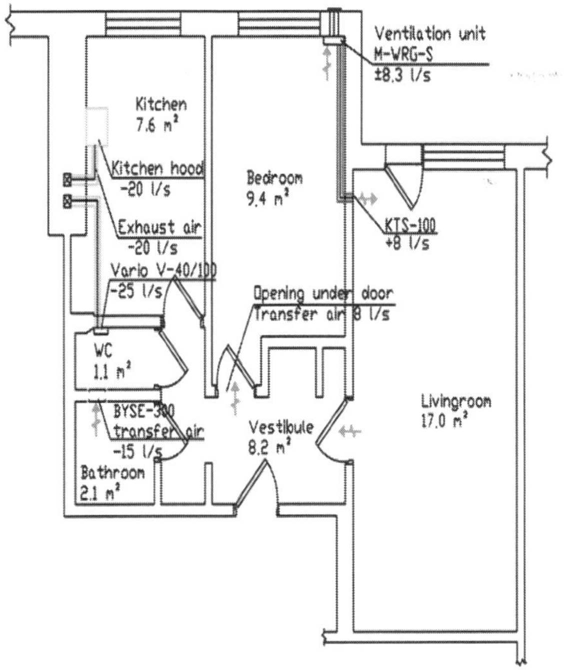

Figure 7. Meltem AHU in a 2-room apartment.

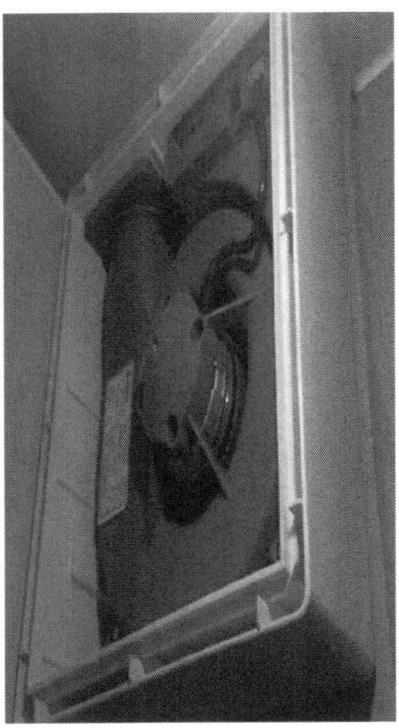

Figure 8. Ventilator in the WC/bathroom, Vario V-40/60.

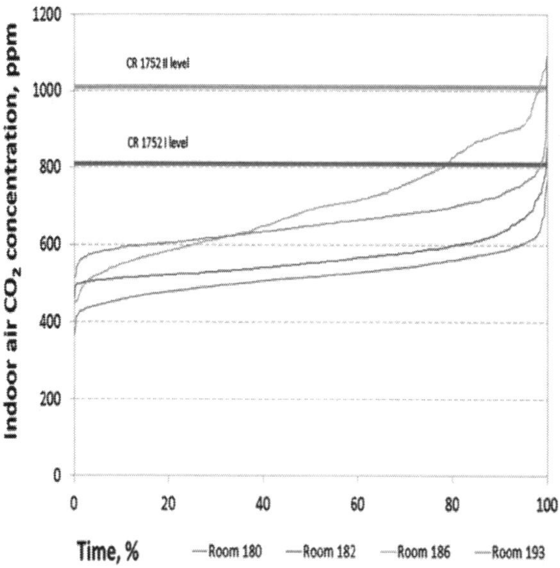

Figure 9. Cumulative distribution of CO_2 concentration in different rooms of the nursing home.

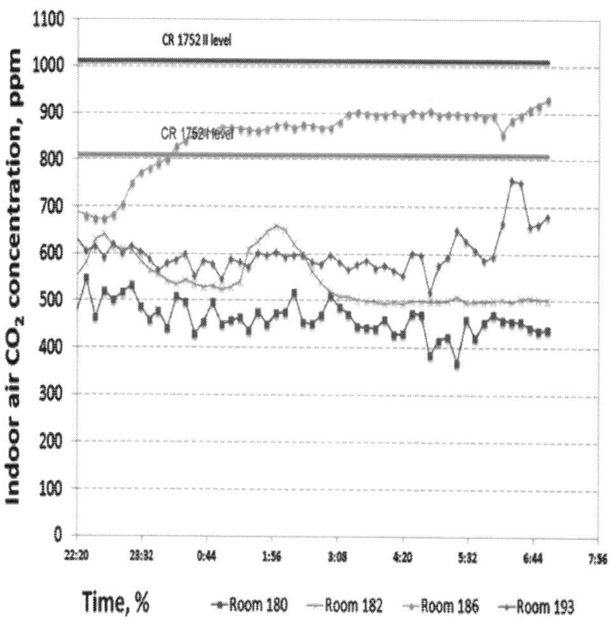

Figure 10. CO_2 concentration in different rooms, air flow rate 25 m^3/h.

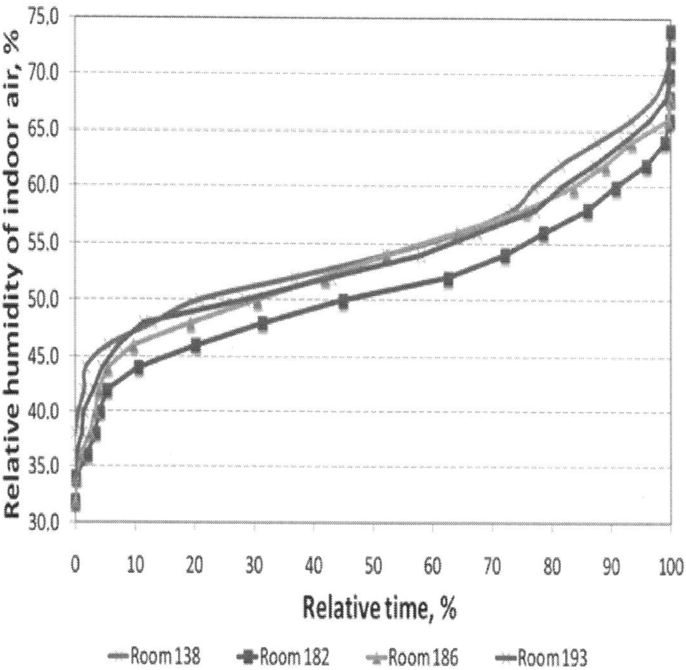

Figure 11. Relative humidity graphs in 4 rooms of the nursing home.

In Figure 12 we see that the concentration of CO_2 in the bedroom after ventilation renovation was less than 1200 ppm, with 70% of the time less than 1000 ppm, which can be considered a good result. The living room had somewhat higher CO_2 levels but more than 90% of the time the concentration was below 1500 ppm.

Tenants' assessment of the use of Meltem AHU in apartments has been very positive. Carbon dioxide levels in the bedroom have dropped nearly 1000 ppm.

An important study in the meaning of energy consumption was the determination of the temperature ratio of Meltem AHU. The measurement results are shown in Figures 14-16.

The temperature ratios for Meltem AHU with a wide range of external temperature are shown inFigure 14.

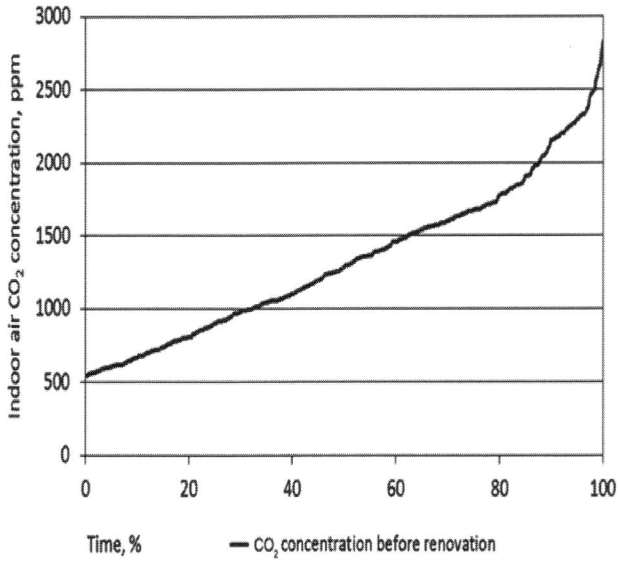

Figure 12. CO_2 concentration in the indoor air before renovating the ventilation (apartement building).

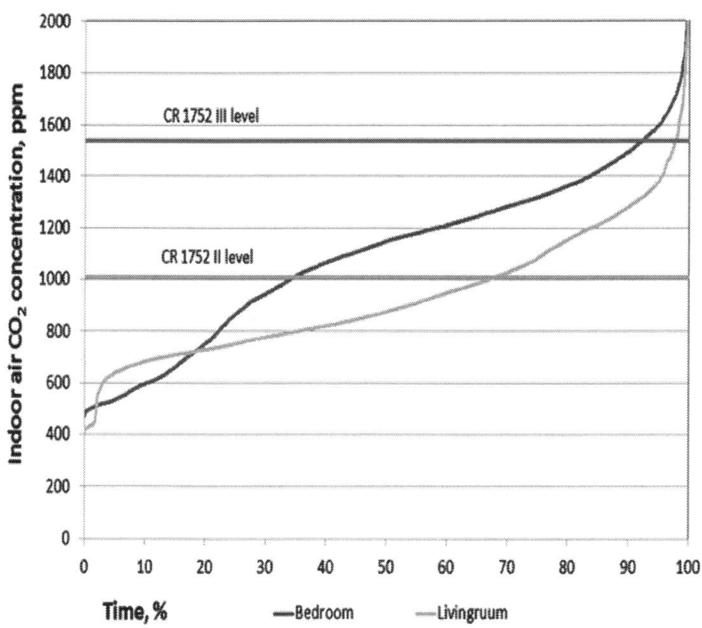

Figure 13. CO_2 concentration after the installation of Meltem AHU.

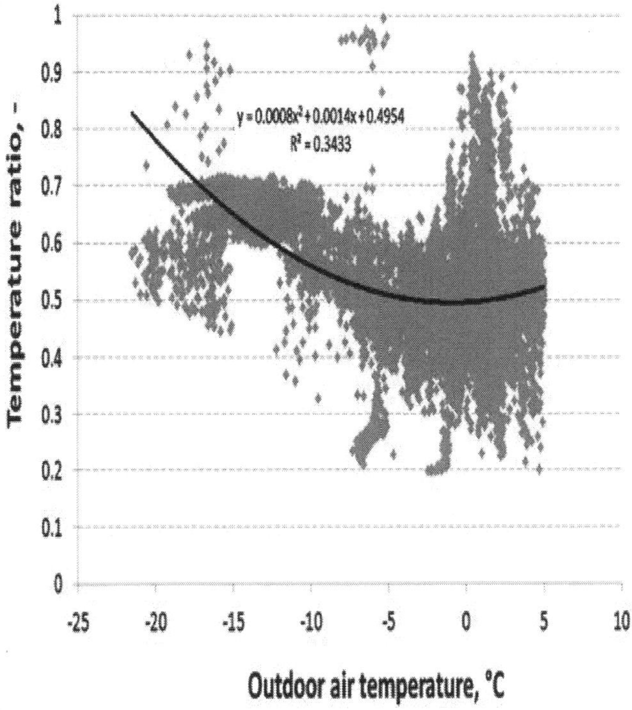

Figure 14. Temperature ratios for Meltem AHU with a wide range of external temperature.

In testing Meltem AHU in the range of outdoor air temperature from +5˚C to −20˚C, the average temperature ratio (efficiency) was between 40% to 65%, with a medium size of 60%. In Figure 15and Figure 16 we can see temperature ratios of Meltem AHU with the 3rd and 5th speed in positive outdoor temperatures. We can see that both temperature ratios are close, with minimum superiority to the fifth speed.

We can see that in real operating conditions the temperature ratio of Meltem AHU is on average close to 0.6, which can be considered a good result.

It should be noted that in old apartment buildings with natural ventilation the use of Meltem AHU in renovating the ventilation is one of the few suitable solutions.

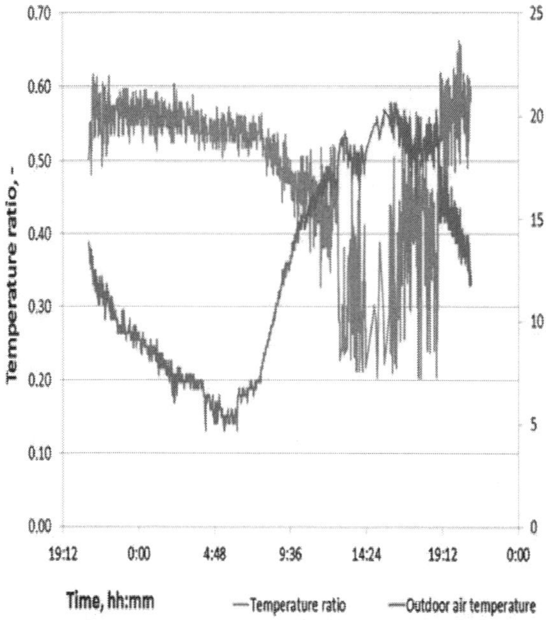

Figure 15. Temperature ratios and outdoor air temperatures for Meltem AHU at the 5th speed.

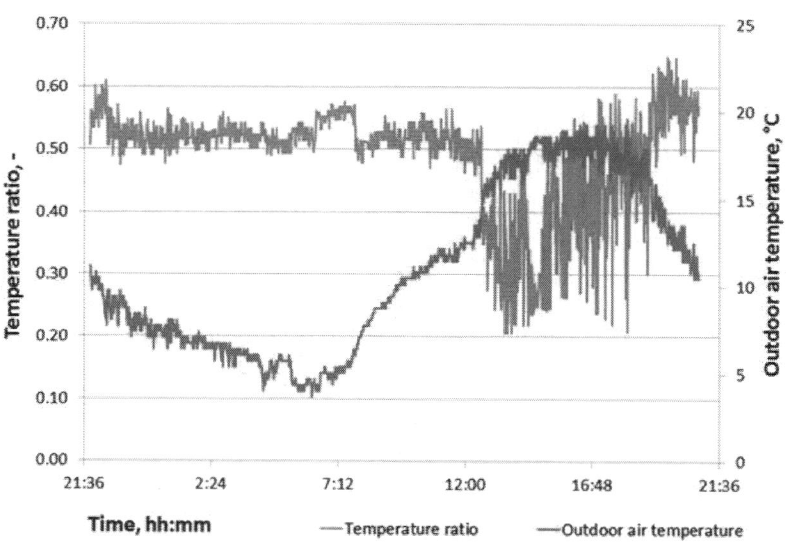

Figure 16. Temperature ratios and outdoor air temperatures for Meltem AHU at the 3rd speed.

Table 4. Noise levels of Meltem AHU-the average of 30 s (Komfort device).

Speed	Air flow rate m^3/h	Noise level in the centre of living room dB (A)	Noise level in the centre of children room dB (A)	Noise level in the centre of bedroom dB (A)
Backround noise	0	22	24.6	23.7
1	15	22.9	25.2	26
2	30	24.1	25.8	26.4
3	60	32.3	34.8	34.8

3.2. Control of the Noise Level

Test measurements of the noise level of Meltem AHU have been carried out in 2-room apartments. The measurements were carried out at 5 flow rates next to the device, specifically 0.3 m from the device. Meltem AHU devices were installed in the premises of 10 m^2, where people sleep.

According to existing normative documents, the noise level should not exceed 30 dB(A) in the bedroom at night and 40 dB(A) in the living room during the day. For measuring the device Delta OHM HD 2010 was used.

The measurement results can be evaluated on the basis of the data given in Table 4. It is seen that Komfort AHU can be used at night at the air flow rate of 15 m^3/h, and with certain reservations at the air flow rate of 30 m^3/h. If someone sleeps in the next room, the device can be used at the air flow rate of 30 m^3/h.

4. CONCLUSIONS

The study shows that the new ventilation solution based on Meltem AHU is perfectly suitable for ventilating rooms in nursing homes and apartments. In the nursing home, Meltem AHU could provide good indoor air conditions in the rooms where two persons live and very good results in the rooms with one person. Measuring results conducted in different rooms show that the room air handling unit with heat recovery ensures good indoor climate in the nursing home. In apartment buildings, people's assessment of the use of Meltem AHU in apartments was very positive. The same studies showed that the temperature ratio of AHU was on average 60% over the season and the quality of indoor environment was satisfactory. Investigation proved that with rational and reasonable organization of ventilation we can save energy in renovating the ventilation in old apartment buildings.

Investments on ventilation with room air handling unit Meltem are close to those with the central balanced ventilation system, for example, in the nursing home.

The running costs of ventilation with the room air handling unit are lower than that with the central balanced ventilation system. The study shows that skilful renovation of ventilation in old apartment buildings enables to get satisfactory indoor climate and energy saving at the same time.

ACKNOWLEDGEMENTS

The research was supported by the Estonian Research Council, with Institutional research funding grant IUT1-15 and with the project "Development of efficient technologies for the air change and ventilation necessary for the increase in the energy efficiency of buildings, AR12045", financed by SA Archimedes. The publication of this article was supported by the ESF measure 1.2.4 Development of cooperation and innovation in universities, the sub-measure "Doctoral Schools" that finances the project "Construction and Environmental Engineering PhD School"—project code 1.2.0401.09-0080.

REFERENCES

1. Frontczak, M. and Wargocki, P. (2011) Literature Survey on How Different Factors Influence Human Comfort in Indoor Environments. Building and Environment, 46, 922-937.

2. Maier, T., Krzaczek, M. and Tejchman, J. (2009) Comparison of Physical Performances of the Ventilation Systems in Low-Energy Residential Houses. Energy and Buildings, 41, 337-353.

3. Jaber, S. and Ajib, S. (2012) Energy Recovery System in Mediterranean Region. Sustainable Cities and Society, 3, 1- 66.

4. Laverge, J. and Janssens, A. (2012) Heat Recovery Ventilation Operation Traded off against Natural and Simple Exhaust Ventilation in Europe by Primary Energy Factor, Carbon Dioxide Emission, Household Consumer Price and Exergy. Energy and Buildings,50, 315-323.

5. Dodoo, A., Gustavsson, L. and Sathre, R. (2000) Primary Energy Implications of Ventilation Heat Recovery in Residential Buildings. Energy and Buildings, 31, 37-47.

6. Koiv, T.-A., Mikola, A. and Kuusk, K. (2012) Energy Efficiency and Indoor Climate of Apartment Buildings in Estonia. International Journal of Energy Science, 2, 94-99.

7. Koiv, T.-A., Voll, H., Mikola, A., Kuusk, K. and Maivel, M. (2010) Indoor Climate and Energy Consumption in Residential Buildings in Estonian Climatic Condition. Wseas Transactions on Environment and Development, 6, 247-256.

8. Mikola, A. and Koiv, T.-A. (2011) Indoor Air Quality in Apartment Buildings of Estonia. Computers and Simulation in Modern Science: Selected Papers from WSEAS Conferences, V, 257-261.

9. Koiv, T.-A. (2007) Indoor Climate and Ventilation in Tallinn School Buildings. Proceedings of the Estonian Academy of Sciences: Engineering, 13, 17-25.

10. EVS-EN 15251:2007 (2010) Indoor Environmental Input Parameters for Design and Assessment of Energy Performance of Buildings Addressing Indoor Air Quality, Thermal Environment, Lighting and Acoustics. Estonian Centre for Standardisation.

11. Regulation of the Government of Estonia No. 38. Dwelling Requirements. 26.01.1999 (RT I 1999, 9, 38) In Estonian.

12. EVS 839:2003 (2003) Indoor Climate. Estonian Center of Standards (In Estonian).

13. CR 1752 (1998) Ventilation for Buildings: Design Criteria for the Indoor Environment. European Committee for Standardization, Brussels.

14. Olesen, B.W. (2007) The Philosophy behind EN15251: Indoor Environmental Criteria for Design and Calculation of Energy Performance of Buildings. Energy and Buildings, 39, 740-749.

15. ASHRAE (1993) Handbook of Fundamentals. American Society of Refrigerating and Air Conditioning Engineers, Inc., Atlanta.

16. European Commission (2007) 2020 Vision: Saving Our Energy. Office for Official Publications of the European Communities.http:// books.google.ee/books/about/2020_vision.html?id=ujsoAQAAMAAJ&redir_esc=y

17. Gustavsson, L., Dodoo, A. and Sathre, R. (2011) Impact of Ventilation Heat Recovery on Primary Energy Use of Apartment Buildings Built to Conventional and Passive House Standard. World Renewable Energy Congress, Linköping, May 2011, 8-11.

18. Diemanu, J., Roth, K.W. and Brodrick, J. (2003) Air-to-Air Energy Recovery Heat Exchangers. ASHRAE Journal, 45, 57-58.

19. Fouih, Y.E., Stabat, P., Riviere, P., Hoang, P. and Archambault, V. (2012) Adequacy of Air-to-Air Heat Recovery Ventilation System Applied in Low Energy Buildings. Energy and Buildings, 54, 29-39.

20. Manz, H., Huber, H., Schälin, A., Weber, A., Ferrazzini, M. and Studer, M. (2000) Performance of Single Room Ventilation Units with Recuperative or Regenerative Heat Recovery. Energy and Buildings, 31, 37-47.

21. Dietz, R.N. and Goodrich, R.W. (1995) Measurement of HVAC System Performance and Local Ventilation Using Passive Perfluorocarbon Tracer Technology. Informal Report, BNL-61990, State University of New York, Farmingdale.

22. Guo, L. and Lewis, O.J. (2007) Carbon Dioxide Concentration and Its Application on Estimating the Air Change Rate in Typical Irish Houses. The International Journal of Ventilation, 6, 235-245.

23. Leephakpreeda, T., Thitipatanapong, R., Grittiyachot, T. and Yungchareon, V. (2001) Occupancy-Based Control of Indoor Air Ventilation: A Theoretical and Experimental Study. ScienceAsia, 27, 279-284. Pavlovas, V. (2006) Energy Savings in Existing Swedish Apartment Buildings. Chamlers University of Technology, Göteborg.

24. Jokl, M.V. (1998) Evaluation of Indoor Air Quality Using the Decibel Concept Based on Carbon Dioxide and TVOC. Building and Environment, 35, 677-697.

25. Kalamees, T., Ilomets, S., Arumägi, E., Alev, Ü., Kõiv, T.-A., Mikola, A., Kuusk, K. and Maivel, M. (2011) Indoor Hygrothermal Conditions in Estonian Old Multi-storey Brick Apartment Buildings. The 12th International Conference on Indoor Air Quality and Climate, Austin, 5-10 June 2011, 6 p.

Chapter 10

Integration of evaporative cooling technique with solar chimney to improve indoor thermal environment in the New Assiut City, Egypt

Amr Sayed Hassan Abdallah , Hiroshi Yoshino, Tomonobu Goto, Napoleon Enteria, Magdy M Radwan, M Abdelsamei Eid

Department of Architecture, Assiut University Graduate School of Engineering, Tohoku University

ABSTRACT

Cooling buildings in summer is one of the main environmental problems for architects and occupants in many hot dry countries. The summer temperature during these countries reaches peaks of more than 40°C in some. Mechanical air conditioners can solve the problem, but they put a heavy strain on the electricity consumption. Egypt in general has rich sunny and clear skies. Therefore, these conditions encourage to enhance evaporating with natural ventilation and save energy. This paper develops an integration of direct evaporative cooling tower with a solar chimney multi-zone thermal ventilation model. Simulation is done using commercial couple multi-zone airflow under COMIS-TRNSYS software (Madison, WI, USA) to assess natural ventilation and indoor thermal comfort. The results show that the system generates 130.5 m^3/h under the effect of solar radiation only and minimum 2 ACH without pressure coefficient which is considered the minimum requirement of ACH. The findings show that the new integrated system interacts with the building envelope and weather conditions to achieve a decrease in indoor temperatures that reach 10°C to 11.5°C compared to outdoor temperatures.

KEYWORDS

Passive cooling, Solar chimney, Integration system, TRNSYS-COMIS, Multi-zone models

1. BACKGROUND

Buildings contribute over 40% of the total global primary energy use corresponding to 24% of the world CO_2. Heating, ventilation, and air conditioning (HVAC) system is the technology of indoor and vehicular environmental comfort which is responsible for about half of the energy use in buildings [1]. The quest to reduce environmental impacts of conventional energy resources and, more importantly, to meet the growing energy demand of the global population had motivated considerable research attention in a wide range of environmental and engineering application [2]. Effective integration of passive features into the building design can significantly minimize the air conditioning demand while maintaining thermal comfort [3]. Passive evaporative cooling is one of the most efficient and long recognized ways of inducing thermal comfort in predominantly hot dry climates. Historically, evaporative cooling was used extensively in traditional architecture throughout the world's hot arid countries [4–6]. In order to extend the use of evaporative cooling, short wind tower cooling was integrated with inclined solar chimney to increase the pressure difference and the system efficiency. This was done after observing the indoor discomfort for one of the Egyptian houses built in the New Assiut City [7].

Many researches have been conducted on using natural ventilation and evaporative cooling strategies for producing cool air and the effect of using a solar chimney on thermal-induced ventilation in buildings. Three important researches were under taken to study these passive systems. Bahadori introduced the idea of capturing the wind within a tower and then pass it through wetted conduit walls [4]. Maerefat and Haghighi also put forward a new solar system employing a solar chimney together with an evaporative cooling cavity. The numerical calculation showed that this integrated system with the proper configurations was capable of providing good indoor conditions during daytime in the living room even at a poor solar intensity of 200 W/m^{2}and a high ambient air temperature of 40°C [8]. Alemu et al. developed an integrated model incorporating passive airflow components into a coupled multi-zone ventilation and building thermal model. This model allows an assessment of a combination of passive features such as solar chimney and wind-induced earth-air tunnel for both natural and hybrid ventilation systems at the design stage [9]. A lot of awareness is ongoing worldwide on solar energy utilization [10].

The main objective of this paper is to develop an integrated solar chimney with passive cooling strategies into a coupled multi-zone ventilation and building thermal model. This was done by developing a mathematical model to incorporate passive features into a multi-zone ventilation model incorporating a simple thermal model using COMIS-TRNSYS software (Madison, WI, USA). This development system is used in a room of a single zone to study the performances and the advantages of an integrated system in the numerical model. Also, it focuses on the simulated performances and its impact on the comfort of occupants for the summer period with an integrated proposed system.

1.1 A theoretical study of the direct evaporative wet medium

An evaporative cooler (EC) is a device that cools air by passing it across a wet medium. As hot air passes over the wet medium, some of its energy goes to evaporate water from the medium. The air comes out of the other side of the device having given up some of its energy and consequently exits at a lower temperature with higher humidity and constant enthalpy. The minimum temperature that can be reached is the thermodynamic wet bulb temperature (TWBT) of the incoming air. It was reported that half of the conventional cooling device in the south west of the USA was replaced by simple direct evaporative air coolers; this would save 18 million barrels of oil per year [5, 11].

In this paper, heat and mass transfer from the wet medium into the airflow are taken into account. Based on the theoretical study and initial system conditions, evaporative cooler performance is based on the concept of the adiabatic process. This means no heat is added to or removed from the system; the process of exchanging the sensible heat of the air for latent heat of evaporation from the water is an adiabatic process [12].

1.2 A description of the system

The performance of the system is to provide desired comfortable conditions and a suitable rate of ventilation, depending on several parameters such as the ambient conditions (temperature, solar radiation, relative humidity, wind speed, and pressure coefficient). The following dimensions and specifications are applied to the model: The system is located in the New Assiut City, Egypt. New Assiut City is located at 27.30°N latitude and 31.15°E longitude. The solar chimney is oriented towards the south. The calculations were carried out for a single zone and to study the integration of the new proposed system, having a dimension of 4.0 m × 4.0 m × 3.125 m. This system was developed from the reference system of inclined solar chimney attached to a room with an evaporative cooling cavity (ECC) [8]. The old model assumes zero pressure boundaries at the inlet and outlet except for the stack effect. Therefore, it is not taken under consideration the variety of the buildings' static pressure from the atmospheric pressure and the effect of wind pressure coefficient on buildings. Also, there are many reasons that make it very difficult to integrate this system into Egyptian housing. The ECC strategies consumed a lot of water because of water film and a lot of space. Figure 1 shows the diagram of the reference system and the new proposed system.

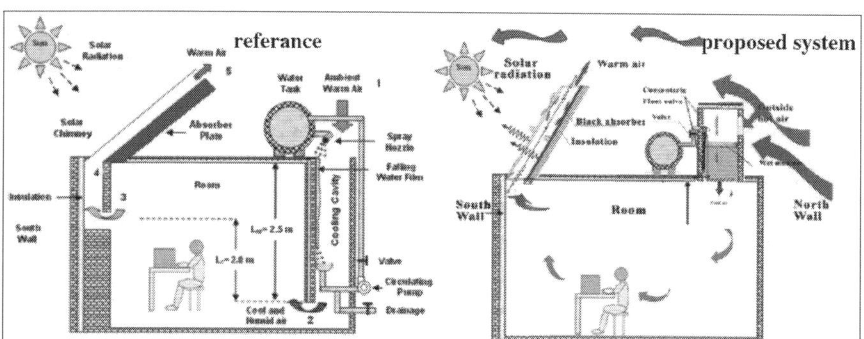

Figure 1. Diagram of solar chimney with evaporative cooler in the reference system and new proposed system.

Nevertheless with the new proposed system, the inclination length of the chimney is 2 m on the inclined angle in order not to extend the vertical height (1.5 m) according to Egyptian Building Regulation Law, and the height of the cooling tower is 1 m. The chimney air gap is 0.2 m, and the inclination of the chimney is 50° based on the angle of the reference system. Table 1 shows the parameters of the solar chimney. Figure 2 shows the location of the room and its dimension in relation to a northern orientation. The performance of the system was monitored at a steady state of condition and then 5 days in the month of June (19, 20, 21, and 23) were selected for studying the performance of the solar chimney between sunrise and sunset in order to study the time-wise to focus on the hottest day in the summer season (20 June). Then, parametric studies for solar chimney and wind tower are needed in the future studies in order to choose the best dimensions for a new proposed compact design. Some parameters used in the calculations of solar chimney appear in Table 1.

Table 1. The parameters used in the calculation**[7]**

Parameters	Value (dimensionless)
Transmissivity of glass	0.84
Absorptivity of glass	0.06
Emissivity of the glass	0.90
Absorptivity of absorber wall	0.95
Emissivity of the absorber surface	0.95

The evaporative pad provides a large water surface, and the pad is wetted by dripping water from the above source. The EC works by a concentric float valve, which opens when the water level is low in the collecting grill, allowing more water to enter. When the water level returns to the full level, the valve is shut automatically.

2. METHODS

The system is being developed with a solar chimney and a small evaporative cooling wind tower. The experimentation would become time consuming and very expensive. Therefore, the numerical model is made and developed for the system. The developed model is implemented in the COMIS-TRNSYS simulation software. This direct evaporative cooling tower used a component 506d-TESS library which uses a wet medium at the top, and this component was assembled and validated by the Thermal Energy System Specialists, LLC (Madison, WI, USA) and was modified in 2004. The present model includes special airflow component which needs simultaneous prediction of temperature and airflow rates. In the developed multi-zone ventilation model, a building is idealized as a system of zones, openings, and ducts linked together by discrete airflow paths. Zones are represented by nodes. Chimney and evaporative cooling channels are represented by ducts for resistance and pressure drop calculations. A hydrostatic condition is assumed in zones, and the flow rate in each link is defined as a function of zone pressure, which results in a system of nonlinear equation solver, defined by the mass conservation for each zone [13]. Figure 3 shows the flow chart solving procedure of TRNSYS-COMIS.

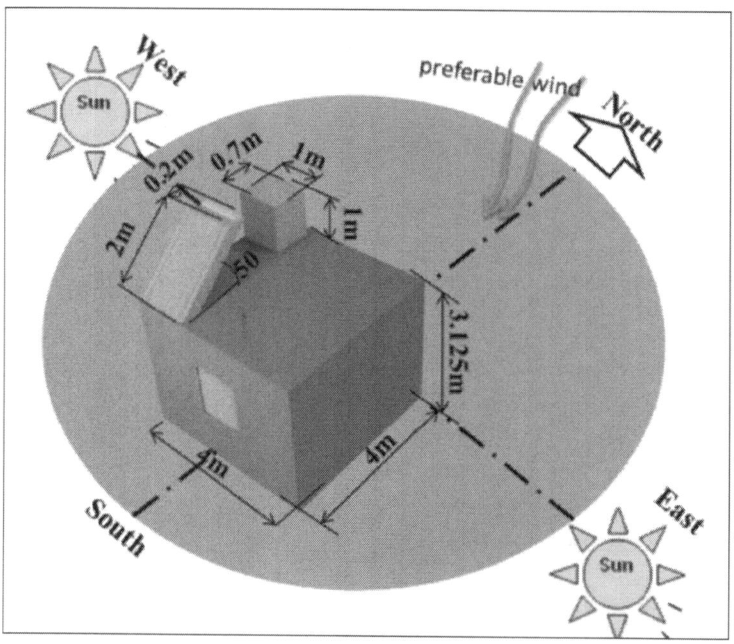

Figure 2. The location of the room with the integrated system according to north orientation.

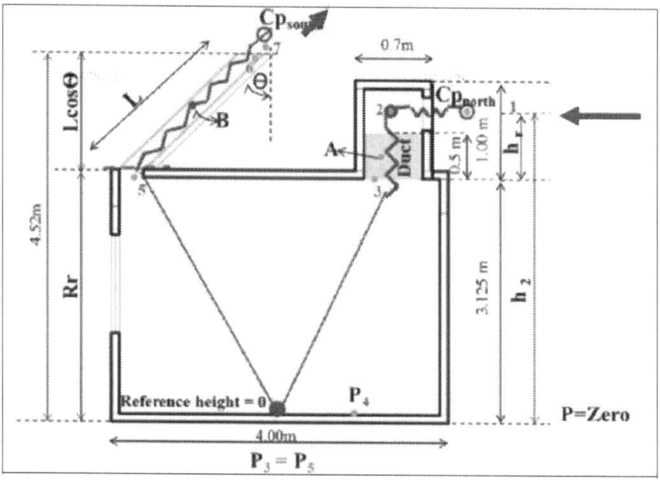

Figure 3. A zone connected to a roof solar chimney with wind tower.

The equations are solved numerically to predict zone pressures, and zone pressures are back substituted in the flow equations to predict the airflow rate at each link [13]. The detailed equations used for large openings and ducts are presented with the fundamentals of the multi-zone airflow model (COMIS, [14, 15]. A description of the mathematical models used in each component can be found in the TRNSYS 16 manual handbook. Multi-zone airflow network models deal with the complexity of flows in a building by recognizing the effects of internal flow restrictions. They require extensive information about flow characteristics and pressure distributions [16]. The pressures are calculated in the previous steps as an initial guess for pressures in the next step. Initial zone densities are based on the pressures found in the last step. The solver updates zone densities as it updates zone pressures [13]. Figure 4 shows the flow chart solving procedure for COMIS-TRNSYS.

2.1 Pressure calculation and mass flow rate through openings

To estimate the amount of air flow through an opening, it is necessary to know the pressure difference across the opening and its effective flow area.

The pressure at an opening can be due to wind, as well as buoyancy, and it is therefore determined by the location of the opening, as well as the internal and external environmental parameters such as wind velocity, pressure coefficient, and outdoor temperatures.

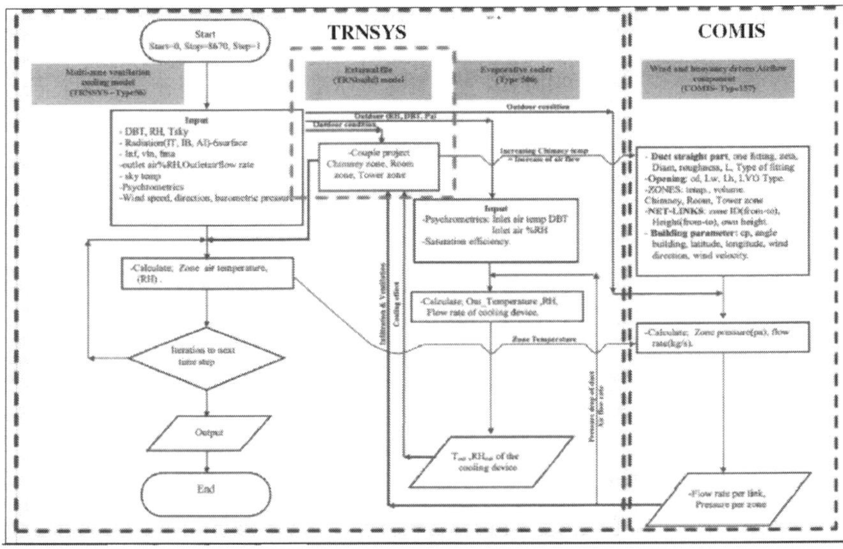

Figure 4. Flow chart solving procedure of TRNSYS-COMIS.

As solar radiation passes through the glazing and absorbed at the wall surface. The air in the chimney is then heated by convection and radiation from the absorber. The decrease in density experienced by the air causes it to rise, whereupon it is then replaced by air from below, i.e., from the attached room. The rate at which air is drawn through the room depends upon the buoyancy force experienced (i.e., depending upon the temperature differential), the resistance of flow through the chimney, and the resistance to the entry of fresh air from the room. While in a wind tower, the opening faces north towards the preferable wind. This compact design with dimensions of 1 m (width) × 0.7 m (depth) and a height of 1 m depends on solar radiation and solar chimney that sucks air from the room and increases the air flow inside the room.

The ventilation rate inside the system is calculated according to six variables (P_2, P_3, P_4, P_5, P_6, and P_7) which are solved from Equations 1 to 6, taking into account the base pressure on the ground ($P = 0$) as presented in Figure 3.

$$P_2 = \left(-\rho g h_2 + 0.5\rho_1 Cp_{\text{north}} v^2\right) - \frac{\rho_2}{2}\left(\frac{Q}{\alpha_1 A_1}\right)^2, \tag{1}$$

where $P_1 = (-\rho g h_2 + 0.5\rho_1 Cp_{\text{north}} v^2)$ is the pressure at point 1 N/m^2, α is the discharge coefficient of the opening in the wind tower, and A_1 is the area of the opening (1 m^2).

$$P_3 = P_2 - \left(\zeta_{\text{WetMediumInlet}} + \zeta_{\text{DuctOutlet}} + \zeta_{\text{DuctFriction}}\right)\frac{\rho_3}{2} v_A^2 + \rho_A g h_r, \tag{2}$$

where the local pressure losses in the inlet $\zeta_{WetMediumInlet}$ of the evaporative cooling (wet medium), the friction of duct $\zeta_{DuctFriction}$, and local losses of the duct outlet $\zeta_{DuctOutlet}$ can be calculated according to the dimensions of dynamic losses due to duct fitting, R e, roughness, duct dimension, and λ. It can be determined in details from COMIS [13]. Also, the indoor velocity equals $V_A = \frac{Q}{A}$ (m/s).

$$P_4 = P_3 - \rho_3 g R_r \tag{3}$$

$$P_5 = P_4 + \rho_4 g R_r \tag{4}$$

$$P_6 = P_5 - \left(\zeta_{\text{ChimneyInlet}} + \zeta_{\text{ChimneyOutlet}} + \zeta_{\text{ChimneyFriction}}\right)\frac{\rho_B}{2} v_B^2 + \rho_B g (L \cos\theta), \tag{5}$$

where the local pressure losses $\zeta_{ChimneyInlet}$ in the inlet of a solar chimney, the friction of chimney $\zeta_{ChimneyFriction}$, and local losses of the chimney outlet $\zeta_{ChimneyOutlet}$ can be calculated in a similar approach as used in evaporative cooling. The solar chimney is treated in the COMIS program as a rectangular duct connects to the room from one side and a node with volume that equals to zero from the other side to transfer the calculated temperature of the chimney from TRNSYS and then another rectangular duct to outside.

$$P_7 = P_6 - \frac{\rho_6}{2}\left(\frac{Q}{\alpha_6 A_6}\right)^2, \tag{6}$$

where $P\,7 = -\rho g(R_r L \cos\vartheta) + 0.5\rho_1 Cp_{south} v^2$, and $\rho = \frac{353.25}{T} = \frac{353.25}{\vartheta + 273.15}$ is the air density (kg/m^3). Then, the pressure differences across the opening due to buoyancy and wind are required and solved with each others in order to predict the pressure of the room and ventilation rate Q (m^3/s).

2.2 Temperature prediction in the solar chimney

The average air density in the chimney channel ρ_6 is predicted using the thermal resistance network model in TRNbuild-TRNSYS, based on the heat balances of each element of solar chimney. The model used by Ong [17] ang Bassiouny and Koura [18] modified by Alemu et al. [9] is adopted with some assumptions in order to solve the mathematical model. Flow through the chimney was considered in a steady state. The chimney inlet area and exit areas are equal. The thermal network for the physical model considered is shown in Figure 5.

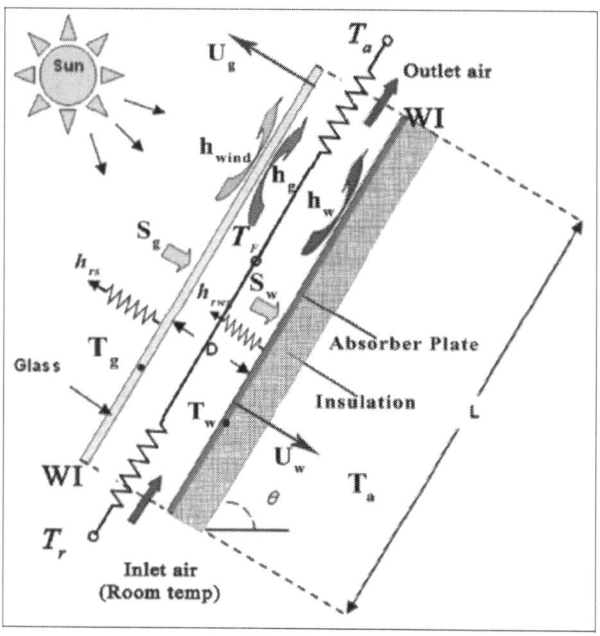

Figure 5. Thermal network for solar chimney with air flow.

The heat balance equations from the thermal network at each element of the solar chimney are described in Equations 7, 8, and 9:

$$T_g(\text{glass cover}):\ S_g + h_{rwg}(T_w - T_g) + h_g(T_f - T_g) = U_g(T_g - T_a) \tag{7}$$

$$T_{f(\text{Air})}:\ h_w(T_w - T_f) = h_g(T_f - T_g) + q'' \tag{8}$$

$$T_w(\text{absorber}):\ S_w = h_w(T_w - T_f) + h_{rwg}(T_w - T_g) + U_w(T_w - T_a). \tag{9}$$

The energy balance on the air flowing through the chimney q'' with a length L and width of the air channel:

$$q^{WL''} = \dot{m}\, C_f(T_{airout} - T_{airin}), \tag{10}$$

where $\dot{m}$ is the mass flow rate of the air component.

The mean air temperature T_f is given by

$$T_{f(air)} = \gamma T_{airout} + (1-\gamma) T_{airin}, \tag{11}$$

where γ (mean temperature weighting factor) = 0.75 [19].

Consider the air inlet to the chimney that has a temperature equal to the room average temperature, substitute in the above equation of q''_{WL}:

$$q''_{WL} = \dot{m}\, C_f \frac{T_{f(air)} - T_{airin}}{\gamma}. \tag{12}$$

The heat transfer coefficient from the glass to atmospheric air

$$U_g = h_{rs} + h_{wind}. \tag{13}$$

The radiation heat transfer coefficient between the glass and the sky is

$$h_{rs} \frac{\sigma_g \varepsilon_g g\left(T_g + T_{sky}\right)\left(T_g^2 + T_{sky}^2\right)\left(T_g - T_{sky}\right)}{T_g - T}, \tag{14}$$

where σ = solar absorptance, ϵ = emissivity, g = gravitational acceleration (9.81), $T_{sky} = 0.0552\, T_a^{1.5}$.

The convective heat transfer outside the glass due to the wind is

$$h_{wind} = 5.7 + 3.8V_{(\text{wind speed at reference height})}. \tag{15}$$

The radiation heat transfer coefficient between wall and glass h_{rwg} is

$$h_{rwg} = \frac{\sigma\left(T_g + T_w\right)\left(T_g^2 + T_w^2\right)}{1/\varepsilon_g + 1/\varepsilon_w - 1}. \tag{16}$$

The heat transfer coefficient between the absorber and outside air is

$$U_w = \frac{1}{1/h_w + \Delta\omega_{insulation}/k_{insulation}}, \tag{17}$$

Where Δ w is the thickness of insulation, $k_{insulation}$ is the thermal conductivity

The convective heat transfer coefficient between air and absorber plate or glass ($h_{air.absorber}$, $h_{air.glass}$)

$$h_{w.x} = \frac{Nu_{air.x} k_{air}}{d_{chimney.gap}}, \tag{18}$$

where x refers to absorber plate or glass, and d is the channel gap.

The Nusselt number correlation for a constant heat flux on one side of the channel [20] is

$$\mathrm{Nu}_{air.x} = 0.9282 Ra_{air.X}^{0.2035}\left(\frac{d_{ChimneyGap}}{Dl}\right)^{0.8972}, \tag{19}$$

where $Ra_{air.x}$ is the Rayleigh number for constant heat flux, Dl is the length.

The solar energy absorbed by the glass is

$$S_{\mathrm{g}} = \sigma_{\mathrm{g}} I_{\text{incident radiation}}. \tag{20}$$

The solar energy absorbed by the absorber

$$S_{\mathrm{w}} = \sigma_{\mathrm{w}} \tau_{\mathrm{g}} I_{\text{incident radiation}}. \tag{21}$$

substituted the above equations into Equations 7, 8, and 9 and the three equations solved simultaneously to give the $T_{\mathrm{g(glass\ cover)}}$, $T_{\mathrm{f(Air)}}$, $T_{\mathrm{w(absorber)}}$.

The mean air temperature $T_{\mathrm{(Air)}}$ in the solar chimney channel is used to determine the average air density $\rho_B$which is then substituted in Equation 5; then, the flow rate in the solar chimney is a function of the chimney temperature.

2.3 Temperature predictions in an evaporative cooling wind tower

An evaporative cooling wind tower (ECWT) is used to let the hot dry air to enter from the outside and is cooled when it passes over a wet medium. When the air passes through the evaporative cooler, the component calls the TRNSYS psychrometrics to fully determine the state of the inlet air. The psychrometrics returns the inlet air wet bulb temperature and enthalpy backs the inlet temperature, pressure, relative humidity, and humidity ratio. The performance of the evaporative cooler is based on the concept of adiabatic process. The wet bulb temperature of the air entering and exiting the device is assumed to be constant, where the minimum outlet temperature equals the inlet wet bulb temperature. Figure 6 shows evaporative cooler schematic.

The energy balance equation expressing the heat removed from the air to evaporate the water is expressed as

$$m_{1a} h_{1\ \text{enthalpy}.\text{AirIn}} = m_{2a} h_{2\ \text{enthalpy}.\text{Airout}} + m_{2\text{vapour}} h_{2\text{vapour}}. \tag{22}$$

The rate of heat transfer from air to water in the wet panel surface, $Q_{sensible}$ (the heat exchange by temperature difference), is given by Δy

$$Q_s = h\,A\,\Delta T_{LM} = mc_p(T_r - T_a), \tag{23}$$

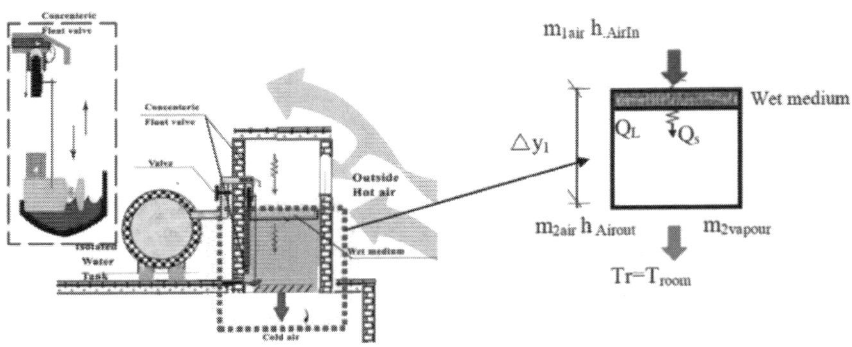

Figure 6. Evaporative cooler schematic.

and the heat gain of latent heat of air (heat exchange by humidity ratio) can be expressed as

$$Q_L = m_L h_L(\Delta w), \tag{24}$$

where *w* is the humidity ratio.

ΔT_{LM} is the log mean temperature difference which is given by [12, 21]

$$\Delta T_{LM} = \frac{(T_r - T_a)}{\ln(T_r - T_{wb})/(T_a - T_{wb})}. \tag{25}$$

The performance of the systems with different flow type can be assessed on the saturation efficiency as

$$\eta\,\text{Saturation} = \frac{T_a - Tr}{T_a - T_{wetbulb}}. \tag{26}$$

The air temperature leaving the cooling device can be calculated from (26)

$$T_r = T_a - \eta\,\text{Saturation}(T_a - T_{wb}). \tag{27}$$

The wet bulb temperature of the air entering and exiting the device is assumed to be constant, where the minimum outlet temperature equals the inlet wet bulb temperature.

Since the performance in the evaporative cooler is an adiabatic process,

$$Q_{\text{Latent}} = Q_{\text{Sensible}}. \tag{28}$$

The humidity ratio of the air outlet from the device [22]

$$\text{RH}_{\text{out}} = \frac{X_{\text{inlet}} p_{\text{outside}}}{(0.622 + X_{\text{inlet}}) p_{\text{saturated.vapour.pressure}}}. \tag{29}$$

The saturate vapor pressure (atm). Bouchahm, 2011 [23] is

$$P_{\text{saturated.vapour.pressure}} = 10^{\left(17.443 - \frac{2795}{T-273} - 3.868 \log_{10}(T-273)\right)}. \tag{30}$$

The corresponding energy removed from the air stream to evaporate water under these circumstances is given by Equation 31:

$$Q_{\text{air}} = \dot{m}_{\text{cooler}}(h_{.\text{AirIn}} - h_{.\text{AirOut}}), \tag{31}$$

where $h_{\text{enthalpy}} = C_p T + x(r + C_{P.\text{vapor}} T)$, expressed in kJ/kg (dry) [24]:

$$1.005\,T + x(2501.1 + 1.846T), \tag{32}$$

where $C_{p.\text{air}}$ = the specific heat of air (kj/kg.K), x = humidity ratio (kg/kg (dry)), r = evaporative latent heat, $C_{p.\text{vapor}}$ = the specific heat of vapor, and T = temperature (°C).

2.4 Air flow network and building thermal model

The ventilation model is integrated to the conventional building thermal model in order to have strong influence on the system performance. The internal heat load, solar load, and the heat transfer through walls, windows, roof, ground, and opening are considered and shown in Figure 7a, where *Q* is the heat transfer. The temperature of the ground at depth of 0.5 m is used in calculation for heat transfer from the ground and derived from Egyptian Typical Meteorological Year (ETMY) which is developed for standards development and energy simulation by US National Climatic Data Center (CA, USA) for a period of 12 years from 1991 until 2003 [25]. Also, the internal load is used in the calculation, where there are four occupants inside the single zone doing light work and seating with light load. Each occupant is assumed to have a sensible gain of 75 W and latent gain of 75 W according to ISO 7730 [26]. The occupants are assumed to be staying in the room for the whole day to apply the maximum load inside. The room light has 13 W/m^2. It has an energy saving lamp with convective part (visible) of 20%, and one device of 140 W is used.

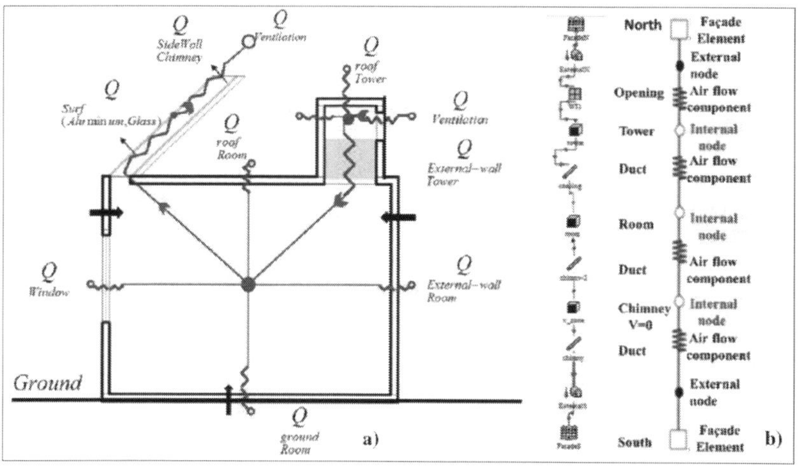

Figure 7. The building thermal model and the air flow network of the integrated model. **(a)** Heat flows in the ECWT, room, and solar chimney with combination of ventilation rate in TRNSYS model; **(b)** air flow network in COMIS with a description of the internal node, external node, and airflow components.

The zones (ECWT, room, and SC) are linked to their ambiance via external nodes and to the north and south facades. Wind pressure data, wind speed, wind velocity profile for the building and for meteo, and orientation are defined. The COMIS program uses wind speed at meteo site in reference height to calculate the speed above the boundary layer using the wind velocity profile of the meteo station. The boundary layer is 60 m high for smooth terrain. Wind pressure coefficients (Cp values) belonging to a certain façade element can be given for several wind directions derived from 'CP-Generator' [13, 27] according to the room with the integrated system with maximum height of 4.5 m and around the building. Figure 7b shows the air flow network in COMIS. Table 2 shows the building construction material description based on the ideal thermophysical properties [28] which was used in TRNBuild. The building thermal model is calculated based on the mathematical Equations 33 to 38.

The heat transfer through external walls (and roof) for room zone and tower zone is predicted as

$$\sum_{\text{external}} Q = \sum^{ne} \frac{A_z(T_{\text{star}} - T_z)}{R}, \tag{33}$$

where ne is the number of external walls attached to the zone, T_{star} is star temperature which can be used to calculate a net radiative and convection heat flux from the wall surface, R is the overall thermal resistance value including convective heat transfer resistance on both sides of the surface, and z is the zone [26].

The total gain to chimney zone from all surfaces is the sum of the combined heat transfers:

$$\sum_{\substack{\text{Surface}\\ \text{Chimney}}} Q = \sum A_s \dot{Q}_{\text{Comb\_}z} + S_{\substack{\text{outside}\\ \text{surface}}}, \tag{34}$$

where A_s is the inside area of the surface, and $S_{\text{outside surface}}$ is the solar radiation for outside surface.

The heat transfer across the windows is

$$\sum_{\text{Window}} Q = \sum A_z U_W (T_a - T_{\text{star}}) + I(\text{SHGF}), \tag{35}$$

Table 2. Building material description used in calculation

Building part	Material part	Conductivity (kJ/h m^{-1} K^{-1})	*U*-value (W/m^2K^{-1})	Thickness (m)
Glass windows	Single glass	-	5.68	0.004
External walls	Common plaster (coating)	1.26	2.60	0.02
	Brick	3.60		0.10
	Common plaster (coating)	1.26		0.02
Roof	Insulation	0.2	0.443	0.05
	Concrete slab	4.2		0.12
	Cement plaster (coating)	4.50		0.01
Ground	Floor	-	0.797	0.10
	Insulation	0.2		0.02
	Concrete	4.2		0.10

where U_w is the overall heat transfer coefficient across the window (the rate of heat transfer through the window from outside to inside when it is hot per unit area and per unit temperature difference (W/m^{2}C), SHGF is the window's solar heat gain factor and its calculation is presented in details [29].

For external surfaces, the long wave radiation is considered using a fictive sky temperature T_{sky}[23]:

$$\dot{Q}_{Comb} = \dot{Q}_{convective} + \dot{Q}_{radiative}$$

$$\dot{Q}_{convective} = h_{ConvOutside}\left(T_{AdjSurface} - T_{outsideSurface}\right) \quad (36\text{ a, b})$$

$$\dot{Q}_{Radiative} = \beta\varepsilon_{outside}\left(T^4_{OutsideSurface} - T^4_{fsky}\right) \quad (36c)$$

$$T_{fsky} = \left(1 - f_{sky}\right)T_{AdjSurface} + f_{sky}T_{sky}, \quad (36d)$$

where f_{sky} is the view factor to the sky, and ϵ is the long wave emissivity of outside surface = 0.9 for walls.

The sky model used to calculate the radiation on inclined surface for the solar chimney by Perez model, which is considered the best available model.

Heat transfer due to ventilation in a zone is

$$\sum_{Ventilation} Q = \sum_{Infiltration} \dot{m}c_p T_{outside} - \sum_{FromZone} \dot{m}c_p T_{zone} + \sum_{ToZone} \dot{m}c_p T_{zoneAdjacment}. \quad (37)$$

Then, the total heat load to maintain the zone at specified temperature can be calculated as

$$Q_{load} = \sum_{Ventilation} Q_{Out} + \sum_{\substack{External\_wall \\ Tower}} Q + \sum_{\substack{External\_wall \\ Room}} Q + \sum_{\substack{ground \\ Room}} Q + \sum_{\substack{Window \\ Room}} Q + \sum_{\substack{Roof \\ Tower}} Q + \sum_{\substack{Roof \\ Room}} Q + \sum_{\substack{Side\_wall \\ Chimney}} Q + \sum_{\substack{Surface \\ Chimney}} Q + \sum_{\substack{People \\ Light}} Q. \quad (38)$$

Then, the zone temperature is iteratively predicted in the calculation.

3. RESULTS AND DISCUSSIONS

3.1 Steady state condition

The performance of the integrated system has been numerically studied, and the effect of the pressure coefficient on ACH and system performance was studied during the steady state condition. The investigation was at constant wind speed at 4 m/s (at Meteo), wind direction of 360°, and pressure coefficient (C_p) 0.081 for the north façade, and -0.040 for the south façade as shown in Table 3. Also, the effect of the different combinations of the three values of solar radiation, temperature, and relative humidity on indoor environment was studied. The main results show the increase of ACH with an increase of solar radiation. The system is capable of generating 130.5 m^3/h ventilation rates for a collector area of 2.4 m^2 under the effect of solar radiation only. The values of these induced air flows depend on the geometry of the air collector and the performance parameters of the solar collector.

Table 3. The results of the numerical simulation after using the new proposed system

Input			Outputs after cooling			
Ambient temperature (°C)	Solar radiation (W/m^2)	Ambient RH (%)	Indoor temperature (°C)	Indoor RH (%)	Indoor ACH	Indoor (m^3/h)
34	1,016	17	34	52.23	6.93	346.84
		25	25.64	58.65	6.95	347.75
38	1,016	17	27.45	51.39	6.80	340.36
		25	28.91	57.97	6.82	341.29
42	1,016	17	30.52	50.63	6.68	334.05
		25	32.13	57.53	6.70	335.09
34	425	17	24.33	52.36	5.45	272.95
		25	25.59	58.82	5.48	274.33
38	425	17	27.41	51.51	5.34	267.05
		25	28.83	58.22	5.37	268.53
42	425	17	30.48	50.74	5.22	261.39
		25	32.08	57.69	5.25	262.97
34	0	17	24.20	52.78	4.28	214.15
		25	25.45	59.31	4.31	215.87
38	0	17	27.28	51.91	4.17	208.75
		25	28.70	58.69	4.21	210.54
42	0	17	30.36	51.11	4.07	203.57
		25	31.95	58.14	4.10	205.45

It can be seen from Figure 8 that the chimney cavity temperature is always higher than the ambient temperature with a very high temperature for an aluminum surface (black absorber) that reaches to 64°C at the radiation level of 1,000 W/m², then the surface temperature decreases to 40°C at the radiation level of 400 W/m². It has been observed that the black absorber possesses a high temperature value compared to chimney air temperature; this increase in temperature is due to more captured radiation and storage of high thermal energy.

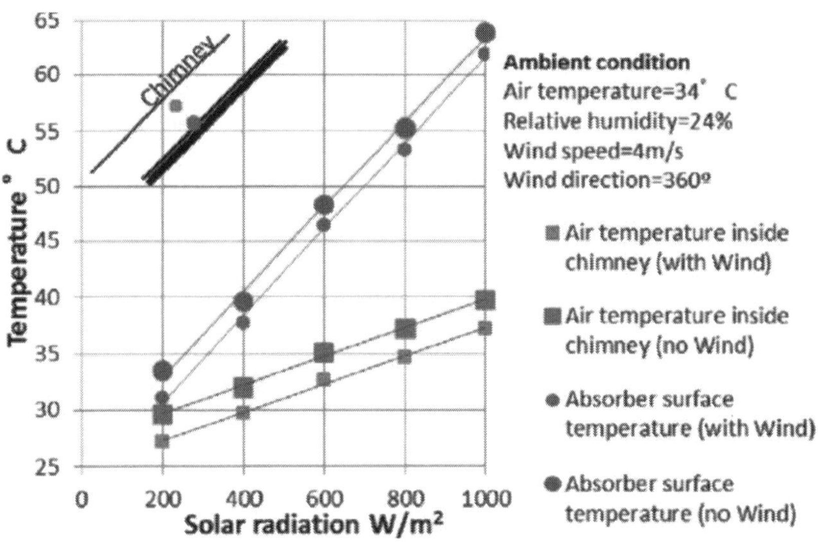

Figure 8. The variation of the absorber and chimney temperature with solar radiation.

This absorbed energy accelerates the flow of air through the chimney during the day and the night. As a result, this exit air flow will help more fresh air to enter to the space from the opening of the wind tower. Figure 9 shows the increase of ACH with an increase of solar radiation and pressure coefficient with maximum 7 ACH inside the room. Also, the minimum indoor air change rate is 2 ACH with 200 W/m² of solar radiation and no wind effect. Thus, the indoor ventilation rate reached zero with no radiation and no wind pressure coefficient. The temperature of the air is higher at the inlet and decreases at the outlet in the evaporative cooler

(EC). This is because, initially, the dry air has a large potential to absorb the moisture and gets saturated as the airflow reaches the outlet of the ECWT.

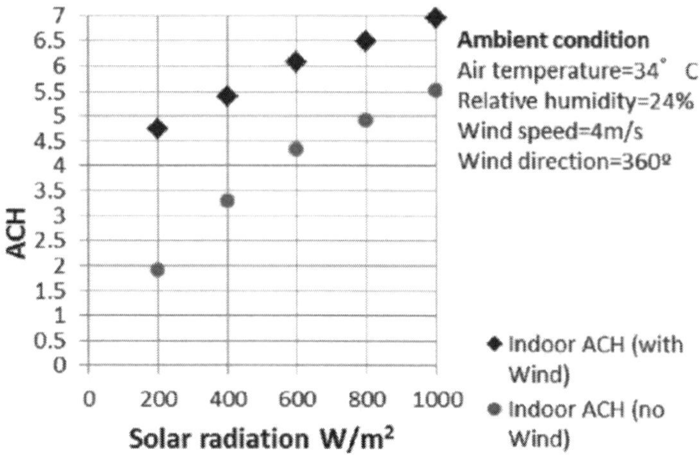

Figure 9. The variation of the ACH with solar radiation with different combinations.

According to ASHRAE 62–2001 standard (Atlanta, GA, USA), ventilation rates dependent on the floor area, whereas the minimum ACH is 0.35, but no less than 15 CFM/person. Also, the Egyptian Energy Code for residential buildings implies a minimum natural ventilation rate of 3 L/s-person (10.8 m^3/h) for the living room [30]. Therefore, the developed system is able to provide the desired ventilation and thermal needs of the occupants with/without solar radiation as shown in Table 3.

3.2 Real weather data

The real weather data of the New Assiut City which is taken from ideal weather data of 20 years [26] was applied to study the system performance during the summer season and the hottest days. In terms of thermal comfort, ASHRAE recognizes that the conditions required for thermal comfort in spaces that are naturally conditioned are not necessarily the same as those required for other indoor spaces [31]. So, a new ACS for a naturally ventilated building has been proposed to be integrated within ASHRAE Standard 55, 2004. In the ACS, the mean monthly outdoor air temperature determines the acceptable indoor operative temperature. This relationship is expressed by the following formula: $To_{(comf)} = 0.31Ta_{(out)} + 17.8$, where $To_{(comf)}$ is the optimum comfort operative temperature in °C and $Ta_{(out)}$ is the mean monthly outdoor air temperature in °C.

Therefore, most of the outlet temperature after passing through the wetted medium in ECWT is below the acceptable limits of Adaptive Comfort Standard (ACS) as shown in Figure 10. Only 5% of the results data exceed the upper limit of 80% acceptable range of ACS especially for the hottest days. An investigation of the system performance during the 5 days of June was done to focus on the hottest day in the summer season from the weather data file so that a clear pattern is seen in relation to adaptive comfort standard (ACS) as shown in Figure 11.

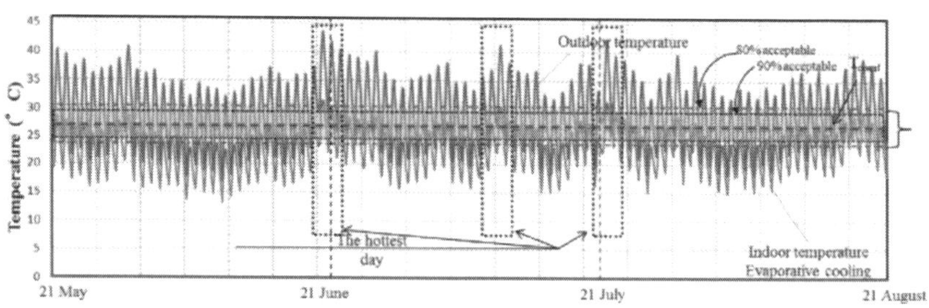

Figure 10. The performance of the system during 3 months of summer using standard weather data.

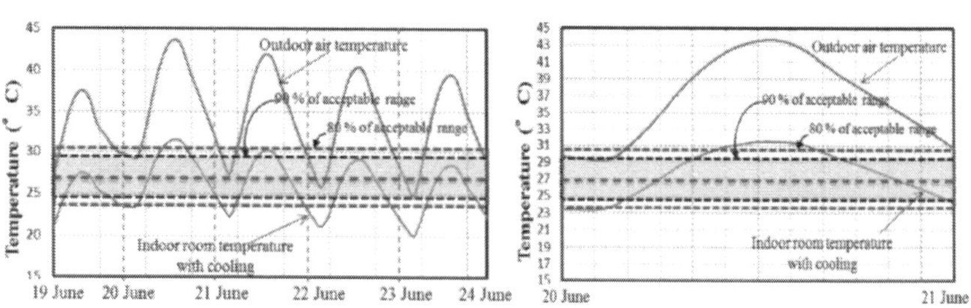

Figure 11. The performance of the system on 19 to 23 June with focus on the hottest day of June.

Arundel et al. proposed that the low or high humidity strongly affects the health and the optimum humidity level which is between 40% and 60% [32]. Therefore, the humidity environment in the living room and the bedroom is very important. Figure 12 shows most of the simulation results for the hottest day, which are located between 40% and 60% except before sunrise between 1 am and 6 am. Thus, this shows the effectiveness of a natural ventilation strategy with a direct evaporative cooling for an indoor environment. Finally, this proposed compact system could be an economical and practical passive alternative to the conventional air conditioning systems in hot and dry climates with no pollution released or energy consumed.

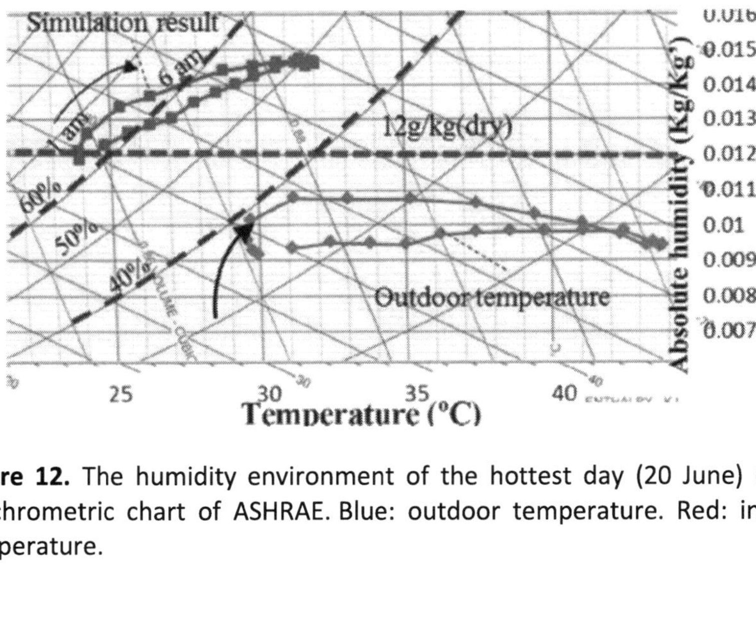

Figure 12. The humidity environment of the hottest day (20 June) using psychrometric chart of ASHRAE. Blue: outdoor temperature. Red: indoor temperature.

4. CONCLUSIONS

Important conclusions are drawn from the numerical operations of a new integrated system. Pressure coefficient and solar radiation in relation to the area of the collector are the main parameters that affect the air flow rate inside a single zone. When the temperature of the black absorber increases with/without pressure coefficient effects, the ACH increases with a strong relation to solar radiation as shown in Figure 8. The temperature of the indoor environment decreases by 10°C to 11.5°C compared to the outdoor temperature. This helps to achieve comfort during the hottest days of the summer season, with 95% of indoor temperature below the upper range of 80% acceptable comfort range during the period from 21 May to 21 August. The proposed system can be applied during daytime and nighttime, but it can be controlled and its use can be limited during nighttime especially when the inlet outdoor temperature is ≤30.5°C.

Overall, the results show how passive systems interact with the building and weather conditions. This paper studied the effect of a new integrated system for indoor comfort for the New Assiut City as a case study. The authors' future work will be a parametric investigation and optimization of the important operating parameter to achieve compact and high performance design so that guidelines for an efficient optimization process will be done to optimize cases with different locations, dimensions, and number of occupants.

Nomenclature

The list of nomenclature is presented in Table 4.

Table 4. Nomenclature

Symbol	Description	Unit
v	Air velocity	m/s
Q	Volume flow	m^3/s
T_{Wb}	Wet bulb depression	°C
Q_{air}	Energy transferred to or removed from the air stream	kJ/h
$\dot{m}_{coolere}$	Mass flow rate of the air through the cooling device	kJ/h
$\dot{m}$	Mass flow rate of the air	kg/s
h_{AirIn}	Enthalpy of air entering the device	J/kg
h_{AirOut}	Enthalpy of air exiting the device	J/kg
T	Temperature	K
t	Temperature	°C
ΔT	Wet bulb depression DRT-WBT	°C
ΔT_{LM}	Log mean temperature difference	°C
C_p	Specific heat at constant pressure	kJ/kg °C
$C_{psouth,north}$	Wind pressure coefficient	-
C_f	Specific heat of air	J/kg K
Δw	Humidity ratio	-

S_g	Solar radiation absorbed by glass	W/m^2
S_w	Solar radiation absorbed by absorber	W/m^2
h_{rwg}	Radiative heat transfer between absorber and the glass	W/m^2 K
h_{rs}	Radiative heat transfer between the glass and the sky	W/m^2 K
h_g	Convective heat transfer between glass and sky	W/m^2 K
h_{wind}	Convective heat transfer outside glass due to wind	W/m^2 K
h_w	Convective heat transfer between the absorber and the air	W/m^2 K
T_a	Temperature of ambient air	K
T_{sky}	Sky temperature	K
T_g	Mean glass temperature	K
T_{fair}	Temperature of air in the chimney	K

ACKNOWLEDGEMENTS

The research leading to these results has received funding from the Egyptian Ministry of Higher Education. The author gives special thanks to Mr. Hans C. Phaff (TNO, The Netherlands), one of the COMIS developers for his scientific help with the simulation software, and to Mr. David Bradley, one of the developers of TRNSYS software.

REFERENCES

1. IEA: Promoting energy efficiency investments. . (2012). Accessed June 2012 http://www.iea.org/publications/freepublications/publication/25recom_2011.pdf

2. Oyedepo SO, Adaramola MS, Paul SS: Analysis of wind speed data and wind energy potential in three selected locations in south-east Nigeria. *Int. J. Energy Env. Eng.* 2012, 3(7):4–10. doi:10.1186/2251–6832–3-7

3. Santamouris M: *Advances in Passive Cooling*. London: Earth Scan; 2007.

4. Bahadori MN: An improved design of wind towers for natural ventilation and passive cooling. *Sol. Ene.*1985, 35(2):119–129. 10.1016/0038-092X(85)90002-7

5. Watt JR: *Evaporative Air Conditioning Handbook*. USA: Chapman and Hall; 1986.

6. Khoukhi M, Fezzioui N: Thermal comfort design of traditional houses in hot dry region of Algeria. *Int. J. Energy Env. Eng.* 2012, 3(5):1–9. Hassan AM, Goto YH, Enteria T, Abdelsamei N, Eid M, Radwan MM: Analysis of thermal comfort for indoor environment of the new Assiut housing in Egypt. *World Acad. Sci. Eng. Tech.* 2013, 77: 845–851. http://waset.org/journals/waset/v77/v77–200.pdf

7. Maerefat M, Haghighi AP: Natural cooling of stand-alone houses using solar chimney and evaporative cooling cavity. *Renew. Ener.* 2010, 35: 2040–2052.

8. Alemu TA, Saman W, Belusko M: A model for integrating passive and low energy airflow components into low rise buildings. *Ener. Build.* 2012, 49: 148–157.

9. Idowu OS, Olarenwaju OM, Ifedayo OI: Determination of optimum tilt angles for solar collectors in low-latitude tropical region. *Int. J. Energy Env. Eng.* 2013, 4: 29. doi:10.1186/2251–6832–4-29 doi:10.1186/2251-6832-4-29

10. Camargo JR, Ebinuma CD, Silveira JL: Experimental performance of a direct evaporative cooler operating during summer in a Brazilian city. *Int. J. Ref.* 2005, 28: 1124–1132.

11. Elmetenania S, Yousfia ML, Merabetia L, Belgrouna Z, Chikouche A: Investigation of an evaporative air cooler using solar energy under Algerian climate. *Ener. Proc.* 2011, 6: 573–582.

12. Energy Performance of Buildings Group: *COMIS Software Version 3.2: Conjunction of Multizone Infiltration Specialists*. Berkeley: Applied Science Division, Lawrence Berkeley Laboratory; 2005.

13. Feustel HE, Raynor AH: *Fundamentals of the Multizone Airflow Model-COMIS*. Berkeley: Annex V, Air Infiltration and Ventilation Center; 1990.

14. Feustel HE, Dieries J: A survey of airflow models for multizone structures. *Energy. Buildings.* 1992, 18: 79–100.

15. ASHRAE: *Handbook of Fundamentals*. Atlanta: American Society of Heating, Refrigerating and Air-Conditioning Engineers; 1985:22.

16. Ong KS: A mathematical model of a solar chimney. *Renew. Ener.* 2003, 28: 1047–1060. Bassiouny R, Koura NSA: An analytical and numerical study of solar chimney use for room natural ventilation. *Ener. Build.* 2008, 40: 865–873.

17. Hirunlabh J, Kongduang W, Nampraki P, Khedari J: Study of natural ventilation of houses by a metallic solar wall under tropical climate. *Renew. Ener.* 1999, 18: 109–119.

18. Pica RG, Volpes R: An experimental investigation on natural convection of air in a vertical channel. *Int. J. Heat. Mass. Transfer.* 2004, 36: 193–208.

19. Bellorio MB, Pimenta JM: *Theoretical analysis of air conditioning by evaporative cooling influence on gas turbine cycles performance*. Ouro Preto: Paper presented at the 18th international conference of Mechanical Engineers; 2005.

20. Kulkarni RK, Rajput SP: Theoretical performance analysis of indirect–direct evaporative cooler in hot and dry climates. *Int. J. Eng. Sci. Tech.* 2011, 3(2):1239–1251.

21. Bouchahm Y, Bourbia F, Belhamri A: Performance analysis and improvement of the use of wind tower in hot dry climate. *Renew. Ener.* 2011, 36: 898–906.

22. Tanaka IT, Terao TH: *Kenchiku Kankyo Kougaku*. 3rd edition. Japan; 2006:55–60. [in Japanese]. ISBN 978–4-7530–1742–3C. http://www.inoueshoin.co.jp/books/ISBN4–7530–1742–7.html [in Japanese]. ISBN 978-4-7530-1742-3C.

23. US Department of Energy: Building Energy Software Tools Directory. (2012). Accessed Sept 2012 http://apps1.eere.energy.gov/buildings/tools_directory/subjects_sub.cfm

24. Klein SA, Beckman WA, Mitchell JW, Duffie EN, Duffie NA, Freeman TL, Mitchell JC, Braun JE, Evans BL, Kummer JP, Urban RE, Fiksel A, Thornton JW, Blair NJ, Williams PM, Bradley DE, McDowell TP, Kummert M, Arias DA: *TRNSYS Version 16: A Transient System Simulation Program + Mathematical Equation*. Madison: Solar Energy Laboratory, University of Wisconsin; 2006.

25. Phaff JC, Knoll B, Gids WF: Cp-Generator: pressure simulation program. (2012). Accessed June 2012http://cpgen.bouw.tno.nl/cp

26. Asan H, Sancakter YS: Effects of Wall's thermophysical properties on time lag and decrement factor.*Ener. Build.* 1998, 28: 159–166.

27. Danny HW, Li Joseph CL: Solar heat gain factors and the implications to building designs in subtropical regions. *Ener. Build.* 2000, 32: 47–55.

28. HBRC: *Egyptian code for energy efficiency improvement in buildings*. HBRC Cairo: ECP306, M.O. Housing; 2005.

29. ASHRAE: *Thermal Environmental Conditions for Human Occupancy*. Atlanta: American Society of Heating Refrigerating and Air-Conditioning Engineers, Inc; 2004.

30. Arundel AV, Sterling EM, Biggin JH, Sterling TD: Indirect health effects of relative humidity indoor environments. *Env. Health Per.* 1986, 65: 351–361.

Chapter 11

The Building Fabric Thermal Performance of Passivhaus Dwellings—Does It Do What It Says on the Tin?

David Johnston [1] and Mark Siddall [2,*]

[1]Centre for the Built Environment (CeBE) Group, Leeds Sustainability Institute, Leeds Beckett University, Leeds LS2 9EN, United Kingdom

[2]LEAP: Lovingly Engineered Architecture Process, Durham DH1 4HU, United Kingdom

ABSTRACT

The Passivhaus (or Passive House) Standard is one of the world's most widely known voluntary energy performance standards. For a dwelling to achieve the Standard and be granted Certification, the building fabric requires careful design and detailing, high levels of thermal insulation, building airtightness, close site supervision and careful workmanship. However, achieving Passivhaus Certification is not a guarantee that the thermal performance of the building fabric as designed will actually be achieved *in situ*. This paper presents the results obtained from measuring the *in situ* whole building heat loss coefficient (HLC) of a small number of Certified Passivhaus case study dwellings. They are located on different sites and constructed using different technologies in the UK. Despite the small and non-random nature of the dwelling sample, the results obtained from the *in situ* measurements revealed that the thermal performance of the building fabric, for all of the dwellings, performed very close to the design predictions. This suggests that in terms of the thermal performance of the building fabric, Passivhaus does exactly what it says on the tin.

KEYWORDS

building fabric; dwellings; coheating; heat loss; *in situ* performance; Passivhaus; Passive House; performance gap; thermal bypass

1. INTRODUCTION

The ambition of the Passivhaus concept is to provide appropriate indoor air quality (IAQ) and thermal comfort with minimum energy demand. The basic premise is to address thermal comfort requirements and to create an environment whereby the performance of the building fabric will be such that it permits space heating using the minimum fresh air supply required in order to maintain hygiene [1].

For the purposes of certification, the Passivhaus Standard requires the Space Heating Demand to be <10 W/m^2 of Treated Floor Area (which equates to roughly <15 kWh/(m^2 a)) and the Primary Energy Demand must be <120 kWh/(m^{2}a). This paper is concerned with factors that contribute to a building satisfying the Space Heating Demand, namely superinsulation and airtightness. The *in situ* performance of Mechanical Ventilation Heat Recovery systems is not a part of this study. Major features of Passivhaus buildings include.

1.1. Superinsulation

Transmission losses are limited though the provision of a high performance building envelope. The *U-value* of the external walls is typically in the range 0.1 to 0.15 W/(m^2 K). When the building fabric is so well insulated, the proportional impact of thermal bridging becomes more pronounced. For this reason, emphasis is placed upon limiting the impact of thermal bridging. The Passivhaus Standard calculates heat loss through the use of external dimensions (unlike the England and Wales where internal dimensions are used). This means there is a tendency to overestimate heat loss providing that a visual inspection confirms the insulation is suitably continuous. The result is that Passivhaus buildings can be said to achieve zero thermal bridging. Where a visual inspection identifies significant thermal bridging, calculations are required. Visual inspection cannot be used to predict heat loss when using internal dimensions.

Windows are also superinsulated. In order to provide appropriate thermal comfort under design conditions, Passivhaus Certified windows are required to have a *U-value* < 0.8 W/m^2K. In order to make use of passive solar gains, windows are required to have a minimum *g-value* of 50%.

1.2. Airtightness

Once the external envelope is superinsulated, a large proportion of the heat loss can arise from background (unintended) ventilation heat losses. For this reason, as well as reasons relating to comfort and moisture control, the Passivhaus Standard requires that air leakage (n50) be limited to <0.6 ach @ 50 Pa. This is roughly 16 times less air leakage than is permitted under the Building Regulations for England and Wales.

1.3. Mechanical Ventilation Heat Recovery

Within the heating season air quality and comfort are maintained, and ventilation heat losses minimized, through the use of heat recovery ventilation.

Typical characteristics of buildings that satisfy the Passivhaus Standard are contained in Table 1 [1].

Table 1. Typical characteristics of Passivhaus buildings.

Requirements for the Passive House Standard	Recommended	Best Practice
Heating energy demand Q_h (kW h/(m²a)) Primary energy demand, PE (kW h/(m²a)) Volume related air leakage at 50 Pa, n50 (h^{-1})	<15 <120 <0.6	⩽10 ~72/0 ⩽0.2
Component or Construction		
Insulation of opaque envelope, *U-value* (W/m² K)	<0.15	0.06
Thermal bridge free construction, *i.e.*, Linear thermal transmittance, *ye* (W/(m K))	 <0.01	 <0
Glazing with low *U-value* and high g-value, *i.e.*, Thermal transmittance, *Ug* (W/(m² K)) Total solar energy transmittance, *g-value* (%)	 <0.8 >50	 0.51 58
Window, thermal bridge free construction, insulated frame, *Uw* (W/(m² K))	<0.8	0.75 (with *Ug* = 0.7 W/(m² K))
Air leakage	<0.6 h^{-1} at 50 Pa	<0.2 h^{-1} at 50 Pa
Heat recovery with Net efficiency, *hHE* (%) Heat loss through casing Internal and external leakages (%)	 >75 <5 W/K <3	 92 <1
Electric energy demand for ventilation (including control), p_{el} (W/(m³/h))	<0.45	0.3
Largely reduced losses in the heating installation (including DHW)		
Energy efficient electric appliances (e.g., highest EU appliance energy-label class)	Class A	<60% of Class A average
Recommended limit for primary energy use for household electricity (part of PE requirement), PE (kW h/(m² a))	<55	27 (assuming European electricity mix)

The CEPHEUS (Cost Efficient Passive Houses as European Standards) project was used to investigate the potential for introducing the Passivhaus Standard to the European market. Evaluation of the houses in these projects primarily focused upon the in-use energy and internal environmental performance of the occupied buildings [2]. Questions regarding other aspects of the *in situ* thermal performance of the building fabric were not addressed. One of the only published sources available are the results obtained from a coheating test undertaken on a dwelling situated in Heusden-Zolder, Belgium which was tested between March and April 2004 (see [3,4]). However, this test suffered from a number of issues, such as uneven distribution of heat within the dwelling, temperature stratification and experimental overheating.

In order to address the lack of published data, this paper presents one of the most comprehensive sets of results available on the *in situ* whole dwelling building fabric performance of Passivhaus dwellings.

2. A UK CONTEXT

Research in the UK has recognised that designs (theoretical practices) once built are not delivering the actual energy savings that have been predicted. One significant cause of this has been attributed to thermal bypass [5,6]. Thermal bypass is described as heat transfer that bypasses the conductive or conductive radiative heat transfer between two regions [7].

In preparation for the construction of the Racecourse Development, Siddall [8] observed that because the Passivhaus Standard only directly addresses one form of convective thermal bypass (airtightness) two remaining forms were unaddressed; namely wind tightness and closed loop bypass.

"Open loop" convection arises when one air mass is replaced by another. In this situation air gaps permit airflow, and thus heat transfer, between two regions. Theoretically, poor wind tightness need not result in infiltration of the air barrier. In the case of wind tightness, Deseyve and Bednar [9] established that in one property the *U-value* fluctuated by 660% as a result of bulk air movement through the insulation. This impacts upon thermal comfort and energy demand. Uvsløkk [10] examined the impact of wind tightness upon thermal performance and proposed a wind tightness standard whereby the infiltration of external air should be limited so that the *in situ U-value* does not exceed the notional *U-value* by more than 5%.

"Closed loop" convection may occur when temperature differences exist at the boundaries and air gaps permit re-circulatory air flow. This phenomenon may not always contribute to the net exchange of indoor air with the outside, *i.e.* it does not constitute air infiltration. With regard to closed loop bypass, Lecompte [11] used cavity wall construction (notional*U-value* was 0.34 W/m^2 K) to demonstrate that heat loss increased by 193% when there was a 10 mm gap between the insulation and the warm side of the internal leaf of blockwork, a 40 mm gap between the insulation and the external masonry and gaps at the head and foot of the insulation. Under similar conditions, it was also established that a 3 mm gap resulted in a 158% increase in heat loss. As the notional (design) *U-value* is improved, the proportional impact of these air gaps can be expected to increase.

Research reported by Wingfield *et al.* [12] raised concerns about significant thermal bypass occurring at the party wall (the measured party wall *U-value* was roughly 0.6 W/m^2 K; compared to the regulatory assumption of 0 W/m^2 K). An experimental remedial measure using a cavity sock resulted in an amended party wall *U-value* of 0.1 to 0.2 W/m^2 K. Clearly, in the context of Passivhaus dwellings (or the performance gap), this extent of heat loss remains unacceptable. For this reason, at the Racecourse Development, Siddall experimented with using a membrane to close the party wall cavity [8].

Based upon the above, it can be appreciated that these heat loss mechanisms do not constitute air infiltration. This means that pressurisation tests, conducted in accordance with BS EN 13829, are an unsuitable means of determining the acceptability of performance.

Protokollband 18 [13] forms the basis for the quality assurance requirements of Passivhaus buildings. The Protokollband discusses factors often associated with convective thermal bypass, however, it specifically addresses external wall insulation technology and does not consider party wall conditions, timber-frame or cavity wall systems, all of which are common in the UK. (It should also be noted that Protokollband 18 is only available in the German language).

The Passivhaus Planning Package (PHPP) is the principle design tool for Passivhaus buildings. As a steady-state predictive model, it is unable to consider the full implications of thermal bypass mechanisms.

Within the context of the UK construction industry, two questions arise:

(1)Without directly addressing thermal bypass on all counts, can the Passivhaus Standard provide a credible level of Quality Assurance?

(2)By making efforts to address heat loss arising from the three modes of thermal bypass, as well as conduction, is it possible to close the energy performance gap?

3. METHODS

In order to assess whether the thermal performance of the building fabric as-designed has been achieved in practice, this paper has compared the steady-state predicted heat loss of a small number of Certified Passivhaus dwellings located in the UK with their corresponding *in situ* measured heat loss.

The steady-state predicted heat loss associated with each of the Passivhaus Certified dwellings has been obtained from the Passive House Planning Package (PHPP) predictions. PHPP is a planning tool that enables a number of energy balance calculations to be undertaken for a proposed building. As part of the Passivhaus Certification process, it is a mandatory requirement that each Passivhaus dwelling is modelled and verified using PHPP.

The measured *in situ* heat loss of each of the Passivhaus Certified dwellings has been obtained by undertaking an electric coheating test (see [14,15,16,17]. An electric coheating test is quasi steady-state aggregate test method that has been developed to measure the whole dwelling heat loss (both transmission and infiltration) attributable to an unoccupied dwelling *in situ*. It is termed a "quasi-steady-state" or a "quasi-static" test method as the internal conditions within the tested dwelling are artificially held at a predetermined static state, whilst the external conditions are allowed to vary naturally (dynamically) in response to the external climatic conditions.

Alternatives to the quasi steady-state electric coheating test do exist. A small number of dynamic aggregate test methods have been devised. These include: ISABELE [18], the Primary and Secondary Terms-Analysis and Renormalization (PSTAR) method [19,20,21] and the Quick U-value of Buildings (QUB) method [22]. Although the dynamic methods offer the advantage of much shorter test durations than is required by the quasi steady-state coheating test method, the analysis of the data obtained from these test methods tends to be significantly more complex. In addition, the results obtained from using the PSTAR method have been questioned [23] and the ISABELE and QUB methods are currently under development, so their use has been primarily limited to research and development purposes. Consequently, the only method that has been utilised in any significant numbers in the UK is the electric coheating test method.

In 2010, electric coheating became recognised as an established test method in the UK when the protocol developed by Wingfield *et al.* [17] was incorporated within the Post Construction and Initial Occupation studies undertaken under the Technology Strategy Boards (now Innovate UK's) Building Performance Evaluation Programme [24]. The main aim of this programme was to understand the key factors that influence the in-use performance of buildings [24]. Another advantage associated with undertaking an electric coheating test is that the heat loss figure obtained from the test is directly comparable to the steady-state heat loss figure that is predicted using the appropriate national calculation methodology. In the UK, for domestic buildings, the national calculation methodology is the Government's Standard Assessment Procedure (SAP) [25].

The electric coheating test method involves artificially heating all of the interior spaces contained within the thermal envelope of an unoccupied dwelling to a mean elevated and homogeneous internal temperature. It does so by using strategically positioned, thermostatically controlled, electric resistance point heaters. To ensure that the internal air is appropriately mixed, and to minimise any potential temperature stratification, electrically powered air circulation fans are also strategically positioned within the dwelling (see Figure 1). Once the building fabric is fully heat saturated and is in thermal equilibrium with the indoor conditions, the mean internal temperature is maintained constant for a specified period of time, typically between 7 to 21 days. In the UK, the mean elevated internal temperature is normally set at 25 °C for new build dwellings, although lower temperatures are often used in existing dwellings prior to refurbishment, particularly those that have poor levels of airtightness and thermal insulation.

A mean elevated temperature of 25 °C is often used for a number of reasons. First of all, it ensures that there is sufficient temperature difference between the inside and the outside of the dwelling (ΔT) throughout the test period. Normally, at least a 10 K ΔT or more is recommended. Secondly, it ensures that the direction of the heat flow is predominantly monodirectional from the inside to the outside of the dwelling. By undertaking the tests during the winter heating season (October/November to March/April), monodirectional heat flow is also ensured. Thirdly, it reduces the potential for midday overheating to occur within the dwelling and the subsequent impact that this may have on heat storage effects. Fourthly, it enables the heating season to be artificially extended beyond the period that would normally be experienced within the dwelling. Finally, the adoption of 25 °C as the internal set point temperature also ensures that the mean internal temperature of the dwelling lies within the range of temperatures that would normally be expected to be experienced within the dwelling during occupation. This avoids the occurrence of non-representative heat loss mechanisms and avoids placing any unnecessary thermal stresses on the building that may cause undue degradation of the building fabric.

During the electric coheating test, numerous internal and external parameters are measured. These include: internal and external air temperatures, solar radiation and the total electrical power input to the dwelling. By measuring these various parameters, the total daily heat input to the dwelling in Watts, that is required to obtain a particular ΔT in K, can be established. If the daily heat input (W) is plotted against the daily mean temperature difference between the inside and outside of the dwelling (ΔT), then the raw uncorrected heat loss coefficient (HLC) is determined (W/K) from the resultant gradient of the plotted trend line (see Figure 2). The raw uncorrected heat loss coefficient can then be corrected using multiple linear regression analysis techniques to take into account external environmental effects, such as solar radiation. Further details regarding the analysis techniques that can be used along with a description of the latest version of the electric coheating test method can be found within Johnston *et al.* [26].

4. CASE STUDY DWELLINGS

In the UK, only a small number of Certified Passivhaus dwellings have been the subject of an electric coheating test. In total, tests have been undertaken on seven dwellings across five separate Passivhaus developments. On two of these developments, the Racecourse Development and Future Works, two separate dwellings have been tested.

All of the dwellings, apart from Ford Close, were tested as part of the Technology Strategy Board's Building Performance Evaluation Programme [24].

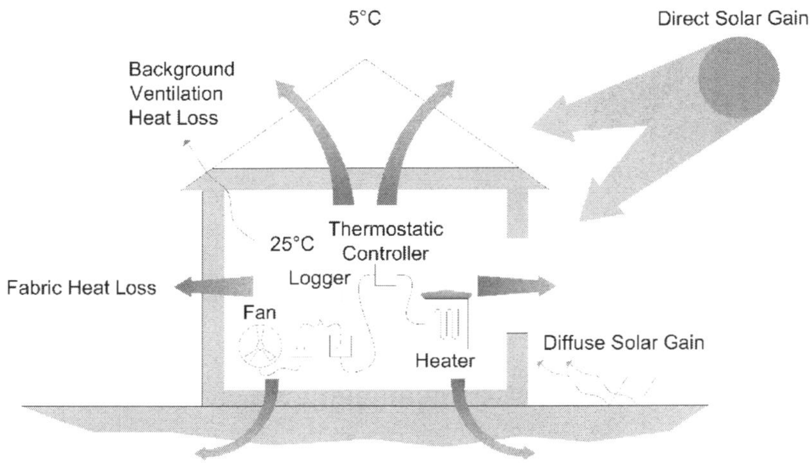

Figure 1. Diagrammatic representation of an electric coheating test [27].

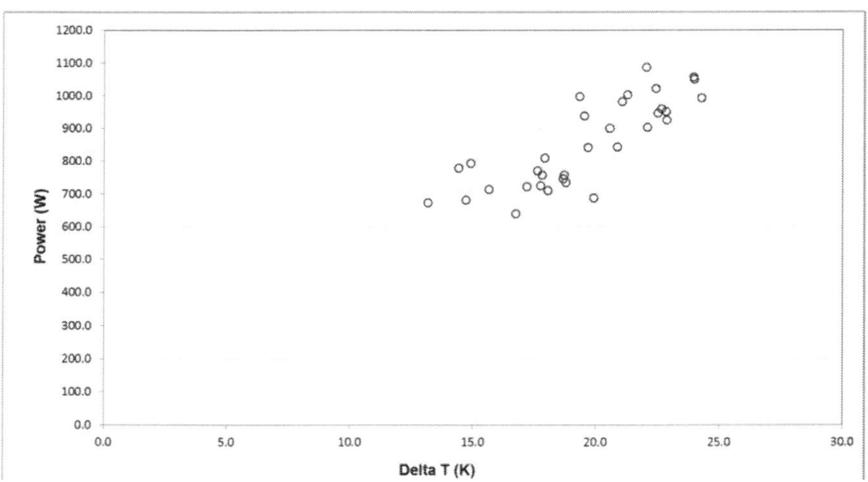

Figure 2. Example plot of the raw uncorrected data obtained from an electric coheating test.

The dwellings that have been tested vary in terms of their size, form and main construction type. In addition, they have been constructed in a diverse range of geographic locations by different clients, designers and construction teams. Details of the individual dwellings and their main form of construction are contained within Table 2 and Table 3.

Table 2. Details of the tested dwellings.

Dwelling	Form	No. of Storeys	Main External Wall Construction	Gross Floor Area (m^2)
Racecourse Development Dwelling 1	End-terrace bungalow	1 plus mezzanine plant area	Pre-fabricated timber-frame cassette	66
Racecourse Development Dwelling 2	Mid-terrace bungalow	1 plus mezzanine plant area	Pre-fabricated timber-frame cassette	66
Lancaster Co-housing	End-terrace	2	Full fill masonry cavity	65
Future Works Passivhaus Larch house	Detached	2	Timber-frame	99
Future Works Passivhaus Lime house	Detached	2	Timber-frame	76
Ford Close	Mid-terrace	2	Pre-fabricated timber-frame cassette	91
Camden Passivhaus	Detached	2	Pre-fabricated timber-frame	118

Further details regarding the dwellings that have been tested can be obtained from Cambridge Architectural Research Ltd. [28,29], Guerra-Santin *et al.* [30], Johnston and Fletcher [31], Johnston and Stevenson [32], Randall Simmonds [33] and Stamp [34].

In terms of the electric coheating tests, all of the electric coheating tests were undertaken during the winter of 2011/12. The specific time periods associated with each test and the total number of days incorporated within the electric coheating analysis are as follows:

- Racecourse Development Dwellings 1 and 2—November/ December 2011, 33 days.
- Future Works Passivhaus Larch house—February/March 2012, 15 days.
- Future Works Passivhaus Lime house—February/March 2012, 18 days.

Table 3. Main form of construction for the tested dwellings.

Dwelling	Ground Floor	External Walls	Roof
Racecourse Development Dwellings 1 and 2	Reinforced concrete ground bearing floor slab, with 300 mm insulation above the slab and 50 mm screed. Design *U-value* of 0.08 W/m^{2}K.	Pre-fabricated timber-frame cassettes filled with 300 mm insulation and clad externally with 15 mm bitroc and brick or render. Internally, a 47 mm insulated service void lined with 25 mm plasterboard. Design *U-value* of 0.10 W/m^{2}K.	Pre-fabricated timber-frame cassette filled with 450 mm insulation and clad in clay roof tiles. Design *U-value* of 0.08 W/m^{2}K.
Lancaster Co-housing	22 mm thick tongue and grooved softwood floating floor over 150 mm reinforced concrete ground bearing floor slab on 250 mm of EPS insulation. Design *U-value* of 0.12 W/m^{2}K.	Two coat wet plaster internal wall finish, 100 mm blockwork, 300 mm cavity fully-filled with mineral wool insulation, 100 mm blockwork and an 8–10 mm render. Design *U-value* of 0.12 W/m^{2}K.	Bobtail trussed rafter pitched roof construction insulated at ceiling level with 500 mm of mineral wool insulation quilt. Design *U-value* of 0.09 W/m^{2}K.
Future Works Passivhaus Larch house	480 mm expanded polystyrene under a concrete ground floor slab and screed. Design *U-value* of 0.07 W/m^{2}K.	Timber-frame with 225 mm insulation between the studs and clad externally with 100 mm rigid wood fibre insulation and clad in larch. Internally, a 100 mm insulated service void lined with 15 mm plasterboard. Design *U-value* of 0.10 W/m^{2}K.	Cold pitched roof with 560 mm glass wool insulation laid horizontally within loft space. Design *U-value* of 0.06 W/m^{2}K.
Future Works Passivhaus Lime house	480 mm expanded polystyrene under a concrete ground floor slab and screed. Design *U-value* of 0.07 W/m^{2}K.	Timber-frame with 225 mm insulation between the studs and clad externally with 100 mm rigid wood fibre insulation and lime render. Internally, a 100 mm insulated service void lined with 15 mm plasterboard. Design *U-value* of 0.10 W/m^{2}K.	Cold pitched roof with 560 mm glass wool insulation laid horizontally within loft space. Design *U-value* of 0.06 W/m^{2}K.
Ford Close	Beam-and-block floor with 150 mm PIR insulation above the floor and screed finish. Design *U-value* of 0.15 W/m^{2}K.	Pre-fabricated timber-frame cassette clad externally with 9 mm wood fibre board and rendered. Design *U-value* of 0.10 W/m^{2}K.	Trussed rafter pitched roof Insulated at rafter level with 500 mm+ blown insulation. Design *U-value* of 0.08 W/m^{2}K.
Camden Passivhaus	Ground bearing floor slab with 380 mm wood fibre insulation in the floor. Design *U-value* of 0.10 W/m^{2}K.	Concrete retaining lower walls to rear and side elevations of dwelling. Pre-fabricated timber framed cassette upper walls filled with 280 mm mineral wool, clad externally in European larch and internally lined with 100 mm wood fibre insulation. Design *U-value* of 0.13 W/m^{2}K retaining walls and 0.12 W/m^{2}K upper walls.	Flat and sloping green roof containing 380 mm mineral wool in sloping section and 280 mm PUR and 120 mm mineral wool in flat section. Design *U-value* of 0.08 W/m^{2}K flat roof and 0.12 W/m^{2}K sloping roof.

- Ford Close—February/March 2012, 11 days.
- Camden Passivhaus—March 2012, 8 days.

5. RESULTS AND DISCUSSION

The electric coheating test results for all seven tested case study dwellings are illustrated in Table 4 and Figure 3. For comparative purposes, the measured heat loss coefficient (HLC) obtained from the coheating test has been compared against the predicted steady-state HLC. The predicted steady-state HLC has been obtained from the PHPP assessments that were submitted for certification purposes and represent the original design intent for each of the case study dwellings. For all of the measured HLC values, the standard error associated with the regression analysis has also been included within the figure. Unfortunately, no standard error has been published for the Ford Close result, so it has not been possible to include this error bar within Figure 3.

Table 4. Coheating Test results for the UK Passivhaus case study dwellings.

Dwelling	Predicted HLC in W/K	Measured HLC in W/K	Difference in HLC in W/K (%)
Racecourse Development Dwelling 1	40.3	46.7 ± 0.5	6.3 (15.7)
Racecourse Development Dwelling 2	35.8	38.1 ± 0.5	2.2 (6.2)
Lancaster Co-housing	40.0	47.3 ± 0.5	7.3 (18.3)
Future Works Passivhaus Larch house	57.6	62.0 ± 4.0	4.4 (7.6)
Future Works Passivhaus Lime house	37.2	45.0 ± 2.0	7.8 (21.0)
Ford Close	45.6	50.4	4.8 (10.5)
Camden Passivhaus	66.0	56.0 ± 5.0	−10.0 (−15.2)

It should be noted that although all of the electric coheating tests have been undertaken in accordance with the 2010 version of the Leeds Beckett University (formerly Leeds Metropolitan University) test method [17], the tests were not all undertaken by the same testing team. Instead, four separate teams undertook the electric coheating tests, all with varying degrees of electric coheating testing experience. For some of the teams, the test represented the first electric coheating test undertaken by the team, whilst at the other end of the spectrum, one of the teams had undertaken in excess of 30 electric coheating tests prior to testing these dwellings. Consequently, there is a potential risk that this inconsistency in testing experience may lead to some uncertainty with regard to the results that have been obtained and their interpretation. This needs to be considered when interpreting the results derived from any comparisons that are made between the test results. These matters notwithstanding, recent work undertaken by the NHBC Foundation [35] which reviewed the coheating test methodology found that broadly similar results were obtained by six different teams, from both academia and industry, when they sequentially tested the same dwelling over a different two week measurement period using their own version of the coheating test method. The HLC measured by the teams varied between 56.7 W/K and 77.1 W/K, compared to a calculated steady-state value of 68.4 W/K. These results suggest that the electric coheating test method is relatively robust.

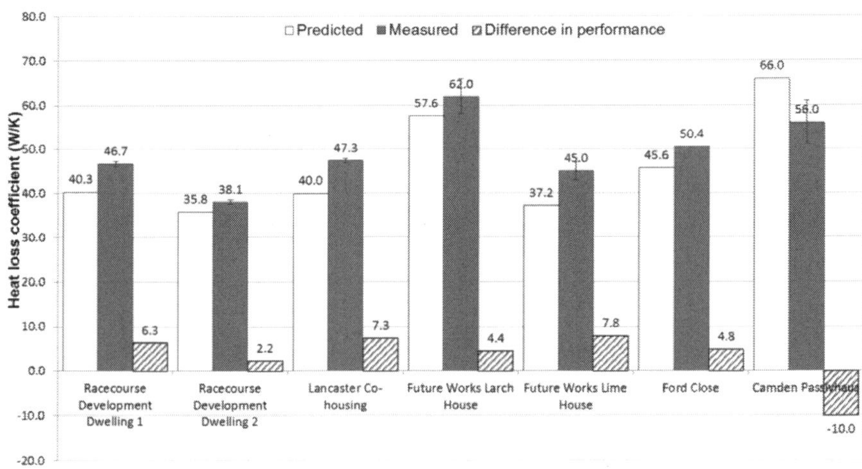

Figure 3. Measured and predicted HLC for each of the case study dwellings.

The results contained within Table 4 and Figure 3 illustrate that not only were the case study dwellings predicted to vary in terms of their building fabric performance, which is understandable given the different sizes and forms of the dwellings, but most importantly, in all but one of the cases (Camden Passivhaus) the measured performance obtained from the electric coheating test exceeded the predicted performance obtained from PHPP.

This indicates the presence of a building fabric "performance gap" in all but one of the Passivhaus Certified case study dwellings.

In terms of the Camden Passivhaus dwelling, it is not entirely clear why this is the only case study dwelling where the measured HLC from the electric coheating test was lower than the predicted HLC. However, closer examination of this case study dwelling reveals that there are a number of factors that may have acted in combination that may have contributed to the result obtained. First of all, the ground floor of the dwelling is partially excavated, resulting in a large proportion of the side and rear ground floor external walls being retaining walls. Consequently, a significant proportion of the dwellings thermal envelope is in direct thermal contact with the surrounding ground, rather than the surrounding air, complicating the analysis of the data. Secondly, the external air temperature sensor was located in a sheltered position for the duration of the test, which may have artificially suppressed the ΔT measured, potentially resulting in an underestimation of the measured HLC. Finally, there may have been some errors in the predicted HLC, particularly with respect to the *U-value* of the retaining external wall, which may have led to a potential "prediction gap". Given the above reasons, the coheating test results obtained for the Camden Passivhaus are considered to be an outlier and have been excluded from any further analysis.

Closer analysis of the data contained within Figure 3 reveals that the size of the gap in percentage terms varies significantly between the dwellings, from just over 6% for Dwelling 1 at the Racecourse Development to just over 20% for the Lime House at Future Works. However, considerable caution should be exercised when using such a metric, as this metric tends to unfairly penalise those that have a very low predicted steady-state HLC to begin with, such as Passivhaus dwellings.

A more appropriate metric for comparative purposes would be to use the absolute difference in HLC between the predicted and the measured performance. Using such a metric results in a very small absolute difference in HLC for all of the tested dwellings, ranging from just over 2.0 W/K for Dwelling 2 at the racecourse Development to just under 8.0 W/K for the Lime House at Future Works.

If one also takes into account the various uncertainties associated with undertaking an electric coheating test, then the size of the building fabric "performance gap" measured for the case study Passivhaus dwellings is very small in absolute terms and is significantly smaller than the size of the building fabric "performance gap" measured in non-Passivhaus dwellings (see Section 6 below). Therefore, from this very limited study, it would appear that the *in situ* performance of the building fabric associated with the case study dwellings is more or less as predicted, suggesting that in terms of building fabric performance, Certified Passivhaus dwellings can in fabric performance terms deliver "what is says on the tin".

6. RESULTS IN CONTEXT

As previously discussed in the introduction, there is a distinct absence of *in situ* building fabric performance data on Passivhaus Certified dwellings, both in the UK and abroad. Therefore, in order to be able to set the results obtained from the case study dwellings in context, the electric coheating test results have been compared against the electric coheating test results of 27 other new build dwellings (see Figure 4 and Figure 5), all of which were originally designed to meet, or in some cases significantly exceed, the building fabric thermal performance requirements contained within Part L1A of the UK Building Regulations 2006 [36]. These results have been obtained from one of the largest and most comprehensive databases of electric coheating data currently available in the UK, the Leeds Beckett Coheating Database (see Table 5).

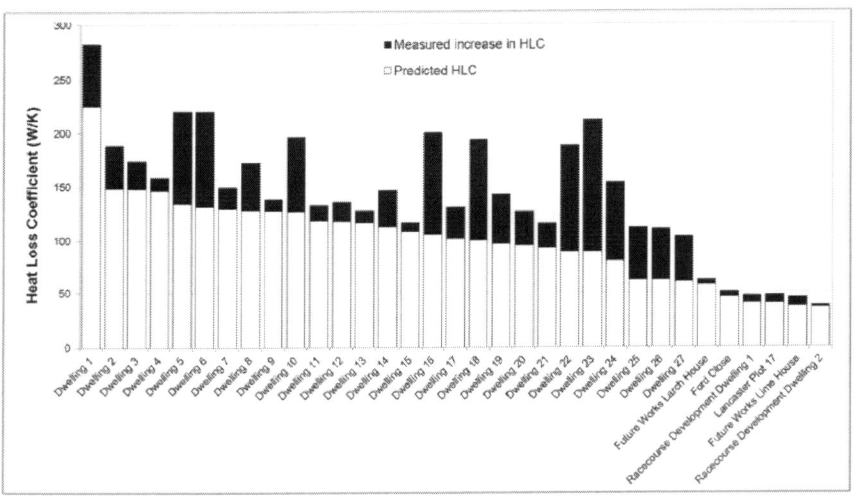

Figure 4. Measured HLC *versus* steady-state predicted HLC of the Passivhaus case study dwellings and the Leeds Beckett new-build coheating database.

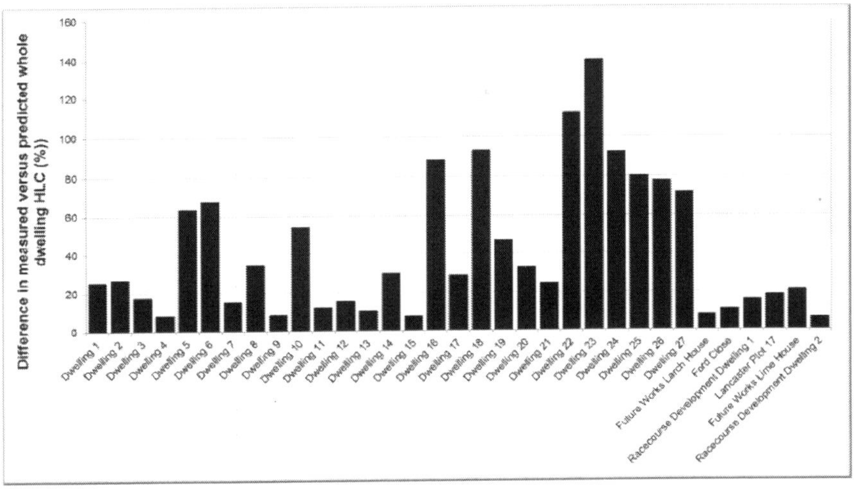

Figure 5. Difference in the measured *versus* steady-state predicted HLCs of the case study dwellings and the Leeds Beckett new-build coheating database.

This database, which represents the compilation of over ten years' experience of undertaking electric coheating tests, contains the results from almost 50 dwellings (in excess of 60 tests), all of which have been tested by the same testing organisation. In a number of instances, the same dwelling has been tested multiple times, enabling the impact of various fabric interventions to be quantified and evaluated. Although the database represents a wide range of both new build and existing dwellings of different size, age, form factor and construction type, the sample size of the database is still small, and the dwellings that are contained within the database are not the result of random sampling. Consequently, the data contained within the database cannot be qualified as being representative of the UK housing stock as a whole. In fact, as a significant proportion of the new build dwellings contained within the database were built to more stringent thermal performance requirements than those that were in force when they were constructed, the new build portion of the database is likely to be biased towards dwellings that had much greater levels of insulation and airtightness than those which would normally have been required for compliance purposes alone. This should be taken into consideration when interpreting the results discussed below. The small number of dwellings contained within the data base also highlights the practical difficulties associated with obtaining aggregate whole dwelling performance data on dwellings in the UK.

Table 5. Size, built form and main construction type of the new build dwellings contained within the Leeds Beckett coheating database.

Dwelling	Main External Wall Construction	Form Factor	Gross Floor Area (m^2)
Dwelling 1	partial fill masonry	detached	167.5
Dwelling 2	full-fill masonry	detached	108.9
Dwelling 3	partial fill masonry	end-terrace	101.0
Dwelling 4	timber-frame	semi-detached	86.7
Dwelling 5	other	detached bungalow	157.0
Dwelling 6	partial fill masonry	end-terrace	102.2
Dwelling 7	thin joint masonry	detached	151.8
Dwelling 8	full-fill masonry	end-terrace	107.5
Dwelling 9	timber-frame	semi-detached	86.4
Dwelling 10	timber-frame SIPS panel	end-terrace	117.0
Dwelling 11	timber-frame SIPS panel	detached	154.5
Dwelling 12	full-fill masonry	end-terrace	94
Dwelling 13	full-fill masonry	end-terrace	94
Dwelling 14	thin joint masonry	end-terrace	141
Dwelling 15	thin joint masonry	end-terrace	141
Dwelling 16	partial fill masonry	end- terrace	83.4
Dwelling 17	thin joint masonry	mid-terrace	141
Dwelling 18	full-fill masonry	end-terrace	141.0
Dwelling 19	thin joint masonry	semi-detached	90.4
Dwelling 20	full-fill masonry	semi-detached	92.6
Dwelling 21	full-fill masonry	semi-detached	92.6
Dwelling 22	full-fill masonry	mid-terrace	137.0
Dwelling 23	partial fill masonry	mid-terrace	83.4
Dwelling 24	full-fill masonry	mid-terrace	106.0
Dwelling 25	full-fill masonry	semi-detached	73.0
Dwelling 26	full-fill masonry	semi-detached	73.0
Dwelling 27	full-fill masonry	semi-detached	73.0

An analysis of the data contained within Figure 4 and Figure 5 indicates that a very wide range of performance exists within the database of new build electric coheating test results, with the measured performance exceeding the steady-state predicted performance in all of the dwellings tested.

What is concerning is that in the majority of the dwellings tested, the difference between the steady-state predicted and the measured performance is significant; in eight of the dwellings the difference is greater than 70% and in two of the dwellings the difference is greater than 100%. On average, the measured fabric performance within the database was just under 45% greater than that predicted. If the Passivhaus Certified case study dwellings are excluded, then the average measured performance difference increases to just over 50% greater than that predicted. Despite this, it is clear that the Passivhaus Certified case study dwellings are the best performing dwellings in the sample, both in terms of predicted and measured performance, by some considerable margin.

In terms of the absolute difference in heat loss, there are also large variations across the database, with the absolute difference ranging from just over 123 W/K for Dwelling 23 to just over 2 W/K for the Racecourse Development Dwelling 2. The average difference in absolute heat loss attributable to the non Passivhaus Certified dwellings is 51 W/K compared to less than 6 W/K for the Passivhaus Certified dwellings. At the most extreme end of the scale, the differences in absolute heat loss measured are likely to have important implications in terms of the energy use and carbon dioxide emissions attributable to these dwellings. Furthermore, the levels of thermal comfort experienced by the occupants of these dwellings is likely to be impaired and it is probable that the performance of the building fabric in these dwellings is in excess of that which is required to achieve Building Regulation compliance. For those dwellings that suffer significant absolute heat loss compared to their prediction (even if they achieve Building Regulations compliance in practice), then with regard to contract law, there may be other legal implications.

7. CONCLUSIONS

The *in situ* thermal performance of the building fabric of seven Passivhaus Certified case study dwellings has been quantified by undertaking an electric coheating test. Although it is only possible to make a number of qualitative comments, due to the small and non-random nature of the case study dwellings tested, the results indicate that in all but one of the dwellings, the measured whole-house *in situ* heat loss was greater than the predicted steady-state heat loss. In the one dwelling where the *in situ* measured heat loss was lower than the predicted steady-state heat loss, it is thought that a number of factors are likely to have acted in combination to contribute to the result obtained.

Although the electric coheating test results indicate the presence of a building fabric thermal "performance gap" in all but one of the Passivhaus Certified case study dwellings, the size of the gap measured was only marginal. If one takes into account the various uncertainties associated with the coheating test, then in spite of the Passivhaus Standard's weak attention to addressing thermal bypass, the building fabric associated with the case study dwellings examined performs within acceptable boundaries and delivers "what is says on the tin". It may be that the knock-on consequence of the onerous airtightness standard required to achieve Certification is that designers and site labour pay greater attention to all aspects of the building fabric which influence building performance.

Regardless of this, the findings of this study can be placed in the context that there is very little measured data available on the *in situ* thermal performance of the building fabric of Passivhaus dwellings. At this point, and with the limited *in situ*data that is currently available, the results suggest it is possible to construct Passivhaus Certified dwellings where the *in situ*performance of the building fabric performs thermally more-or-less as predicted. This is in contrast to the rest of the electric coheating data that are currently available from more mainstream UK housing, where differences between the *in situ*measured and steady-state predicted performance of more than 100% have been observed. Consequently, the Passivhaus Standard would appear to offer a robust solution to the building fabric thermal "performance gap".

ACKNOWLEDGMENTS

The authors gratefully acknowledge the funding provided by the Technology Strategy Board (now Innovate UK) as part of its Building Performance Evaluation Programme (project numbers 450014 and 450095). This funding was used to undertake the electric coheating tests on the three case study dwellings identified within this paper. For further details of these case studies, see Johnston and Fletcher [30] and Johnston and Stevenson [31]. No funding was provided to cover the costs associated with publishing in open access. The authors would also like to thank Jez Wingfield (formerly of Leeds Metropolitan University) who developed an earlier version of the Leeds Metropolitan University (now Leeds Beckett University) coheating database.

AUTHOR CONTRIBUTIONS

David Johnston, along with colleagues from the Centre of the Built Environment Group within the Leeds Sustainability Institute at Leeds Beckett University, performed the electric coheating tests on three of the Passivhaus Certified case study dwellings. The remaining electric coheating test results were obtained from published data. The paper was conceived and written by Professor David Johnston and Mark Siddall.

REFERENCES

1. Feist, W.; Schnieders, J.; Dorer, V.; Haas, A. Re-inventing air heating: Convenient and comfortable within the frame of the Passive House concept. *Energy Build.* **2005**, *37*, 1186–1203.
2. Schnieders, J.; Hermelink, A. CEPHEUS results: Measurements and occupants' satisfaction provide evidence for Passive Houses being an option for sustainable building. *Energy Policy* **2006**, *34*, 151–171.
3. Meulenaer, V.; de Veken, J.; van der Verbeek, G.; Hens, H. Comparison of measurements and simulations of a "passive house". In *Building Simulation 2005*, Proceedings of the 9th International IBPSA Conference, Montreal, QC, Canada, 15–18 August 2005; pp. 769–776.
4. Meulenaer, V.; de Hens, H. Performance Assessment of "Passive Houses" Based on Extensive Measurement. Department of Civil Engineering, Catholic University of Leuven: Leuven, Belgium, 2005.
5. Stafford, A.; Bell, M.; Gorse, C. *Building Confidence—A Working Paper*; Centre for Low Carbon Futures: York, UK, 2012.
6. *Zero Carbon Hub Closing the Performance Gap: Building Low Carbon Housing for Real*; Report of Topic Work Group 4, Carbon Compliance Tool Policy Assumptions Task Group; Zero Carbon Hub: Milton Keynes, UK, 2010.
7. Harrje, D.T.; Dutt, G.S.; Bohac, D.L.; Gadsby, K.J. Documenting air movements and infiltration in multicell buildings using various tracer techniques. *ASHRAE Trans.* **2006**, *91*, 2012–2027.

8. Siddall, M. Thermal bypass: The impact upon performance of natural and forced convection. In Proceedings of the International Passivhaus Conference, Frankfurt, UK, 17–18 April 2009; pp. 339–344.

9. Deseyve, C.; Bednar, T. Increased thermal losses caused by ventilation through compact pitched roof constructions—*In situ* measurements. In Proceedings of the Seventh Nordic Building Physic Simposium, Reykjavík, Iceland, 13–15 June 2005.

10. Uvsløkk, S. The importance of wind barriers for insulated timber frame constructions. *J. Therm. Insul. Build. Envel.***1996**, *20*, 40–62.

11. Lecompte, J. The influence of natural convection on the thermal quality of insulated cavity construction. *Build. Res. Pract.* **1990**, *6*, 349–354.

12. Wingfield, J.; Bell, M.; Bell, J.M.; Miles-Shenton, D.; South, T.; Lowe, R.J. *Evaluating the Impact of an Enhanced Energy Performance Standard on Load-Bearing Masonry Domestic Construction*; Interim Report Number 7; Centre for the Built Environment, Leeds Metropolitan University: Leeds, UK, 2007.

13. Feist, W. *Qualitätssicherung Beim Bau von Passivhäusern*; Protokollband No. 18; Passivhaus Institut: Darmstadt, Germany, 2008.

14. Sonderegger, R.C.; Condon, P.E.; Modera, M.P. *In-situ* measurements of residential energy performance using electric co-heating. *ASHRAE Trans.* **1980**, *86*, LBL-10117.

15. Siviour, J. Experimental thermal calibration of houses. In *Rapid Thermal Calibration of Houses*; Everett, R., Ed.; Technical Report ERG 055 for the Science and Engineering Research Council: Milton Keynes, UK, 1985.

16. Everett, R. *Rapid Thermal Calibration of Houses*; Technical Report ERG 055 for the Science and Engineering Research Council: Milton Keynes, UK, 1985.

17. Wingfield, J.; Johnston, D.; Miles-Shenton, D.; Bell, M. Whole house heat loss test method (Coheating). 2010; Leeds, Leeds Metropolitan University. Available online: http://www.leedsbeckett.ac.uk/as/ cebe/ projects/coheating_test_protocol.pdf (accessed on 3 December 2015).

18. Bouchié, R.; Alzetto, F.; Brun, A.; Boisson, P.; Thebault, S. Short methodologies for *in-situ* assessment of the intrinsic thermal

performance of the building envelope. Available online: https://www.google.com/url?sa=t&rct=j&q=&esrc=s&source=web&cd=1&ved=0ahUKEwj0stCk6qrKAhWF2SwKHZzTAKYQFggcMAA&url=http%3a%2f%2fsustainable-places.eu%2fwp-content%2fuploads%2f2014% 2f10%2fBouchie.pptx&usg=AFQjCNGhe9zx1zKp8FndyE2fGzchgbR5Dw&bvm=bv.111677986,d.bGg&cad=rja (accessed on 12 January 2016).

19. Subbarao, K. PSTAR—Primary and Secondary Terms-Analysis and Renormalization: A Unified Approach to Building and Energy Simulations and Short-Term Testing—A Summary. SERI/TR-254-3347; Solar Energy Research Institute: Golden, CO, USA, 1988.

20. Subbarao, K.; Burch, J.D.; Hancock, C.E.; Lekov, A.; Balcomb, J.D. *Short-Term Energy Monitoring (STEM): Application of the PSTAR Method to a Residence in Fredericksburg, Virginia*; SERI/TR-254-3356; Solar Energy Research Institute: Golden, CO, USA, 1988.

21. Francisco, P.; Siegel, J.; Palmiter, L.; Davis, B. Measuring residential duct efficiency with the short-term coheat test methodology. *Energy Build.* **2006**, *38*, 1076–1083.

22. Mangematin, E.; Pandraud, G.; Roux, D. Quick measurements of energy efficiency of buildings. *C. R. Phys.* **2012**, *13*, 383–390.

23. Palmer, J.; Pane, G.; Bell, M.; Wingfield, J. *Comparing Primary and Secondary Terms Analysis and Re-Normalisation (PSTAR) Test and Co-Heating Test Results*; A Report to the Department for Communities and Local Government; HMSO: London, UK, 2011.

24. *The Technology Strategy Board Building Performance Evaluation, Domestic Buildings—Guidance for Project Execution*; The Technology Strategy Board: Swindon, UK, 2010.

25. BRE. *SAP 2009: The Government's Standard Assessment Procedure for Energy Rating of Dwellings*; 2009 Edition Incorporating RdSAP 2009; Published on Behalf of DECC by BRE; BRE: Garston, Watford, 2011.

26. Johnston, D.; Miles-Shenton, D.; Farmer, D.; Wingfield, J. Whole House Heat Loss Test Method (Coheating). June 2013; 2013 Leeds, Leeds Metropolitan University. Available online: http://www. leedsbeckett.ac.uk/as/cebe/projects/cebe_coheating_test_method_june2013.pdf (a ccessed on 3 December 2015).

27. Brooke-Peat, M.; Leeds Beckett University, Leeds. Personal communication, 2015.

28. Cambridge Architectural Research Ltd. 450023 Camden Passivhaus—Final Report, Domestic Buildings, Phase 1: Post Construction and Early Occupation. A report to Innovate UK as part of the Building Performance Evaluation Programme. 2011; Digital Catapult, London. Available online: http://buildingdataexchange.org.uk/report/Camden+Passivhaus/811/ (accessed on 8 December 2015).

29. Cambridge Architectural Research Ltd. 450019 Future Works Passivhauses—Final Report, Domestic Buildings, Phase 1: Post Construction and Early Occupation. A Report to Innovate UK as Part of the Building Performance Evaluation Programme. 2012; Digital Catapult, London. Available online: http://buildingdataexchange. org.uk/report/Future+Works+Passivhaus/807/ (accessed on 8 December 2015).

30. Guerra-Santin, O.; Tweed, C.; Jenkins, H.; Jiang, S. Monitoring the performance of low energy dwellings: Two UK case studies. *Energy Build.* **2013**, *64*, 32–40.

31. Johnston, D.; Fletcher, M. *TSB BPE Project 450014—Gentoo Passivhaus Development: TSB BPE Phase 1 Final Report*; A Report to the Technology Strategy Board as Part of the Technology Strategy Board's Building Performance Evaluation Programme; Centre for the Built Environment (CeBE), Leeds Metropolitan University: Leeds, UK, 2013.

32. Johnston, D.; Stevenson, F. *TSB BPE Project 450095—Lancaster Co-housing Development: TSB BPE Phase 1 Final Report*; A Report to the Technology Strategy Board as Part of the Technology Strategy Board's Building Performance Evaluation Programme; Centre for the Built Environment (CeBE), Leeds Metropolitan University: Leeds, UK, 2013.

33. Randall Simmonds Passivhaus at Ford Close, St Ive, Cornwall. n.d Randall Simmonds. Available online: http://www.randallsimmonds.co.uk/wp-content/uploads/2012/09/Passivhaus-Brochure.pdf (accessed on 8 December 2015).

34. Stamp, S. Assessing Uncertainty in Co-Heating Test: Calibrating a Whole Building Steady State Heat Loss Measurement Method. Ph.D. Thesis, University College London (UCL), London, UK, 2015.

35. Butler, D.; Dengle, A. Review of Co-Heating Test Methodologies. Report F54; NHBC Foundation: Milton Keynes, UK, 2013.

36. NBS. *The Building Regulations 2000 Approved Document Part L1A: Conservation of Fuel and Power in New Dwellings*; 2006 Edition; NBS: London, UK.

Index